Neurotechnologies of the Self

Jonna Brenninkmeijer

Neurotechnologies of the Self

Mind, Brain and Subjectivity

Jonna Brenninkmeijer
University of Groningen
Groningen, The Netherlands

ISBN 978-1-137-53385-2   ISBN 978-1-137-53386-9   (eBook)
DOI 10.1057/978-1-137-53386-9

Library of Congress Control Number: 2016941103

Cover illustration: Cover image © Marc Woldering

Printed on acid-free paper

This Palgrave Macmillan imprint is published by Springer Nature
The registered company is Macmillan Publishers Ltd. London

Preface and Acknowledgments

This book explores how the use of neurotechnologies to understand or improve the self influences people's subjectivity. The figure on the cover of this book aims to illustrate this effect. When you take a close look at the picture, you will see that the person is made of brainwaves and neurons. When you take a closer look, you will see brain maps (qEEGs: the round heads), and also computer chips. And when you look very closely, you can see family trees and mandalas. This assemblage of brain entities, technical elements, environmental influences, and spiritual accents nicely illustrates the main argument of this book: working on the self by working on the brain does not reduce the self to the brain, but extends the self. Especially in the case of neurofeedback, a brainwave therapy, the self is extended with a brain, and with various physiological, psychological, material, and sometimes spiritual entities that all start working upon the person's feelings, problems, and lives. This 'new self' is not static or fixed, and therefore the figure on the cover has no clear borders. I would like to thank my partner Marc Woldering for designing this cover as well as for all his other forms of support during the periods that I worked on this book.

I conducted the research for this book while working for the 'Theory and History' research group at the University of Groningen. I am very thankful for all ideas, support, and critical comments of my supervisors Maarten Derksen and Trudy Dehue, as well as for all comments and suggestions from my other colleagues: Douwe Draaisma, Stephan Schleim, Jess Cadwallader, Sarah de Rijcke, Adeena Mey, Berend Verhoeff, Felix Schirmann, and Hilde Tjeerdema. I have always appreciated the flexible and inspiring atmosphere of 'TG' and I am glad that – after two research positions elsewhere – I am now part of this group again.

For this book, I studied 'brain work' in many facets. I went to several neurofeedback clinics, interviewed neurofeedback clients and practitioners, was an observer during experiment, and was invited to attend a neurofeedback course, therapy, and some meetings. Without the help of all clients and practitioners, it would have been impossible to write this book. I would especially like to thank Roland Verment who received me several times in his neurofeedback clinic—sometimes with groups of students—and answered many of my questions by e-mail. I would also like to thank the European Neuroscience and Society Network that together with the University of Groningen financed a short-term visit to London to conduct interviews, observations, and archival research.

Parts of this book have already been published in academic journals: an early version of Chap. 4 (*Taking Care of One's Brain*) was published in *History of the Human Sciences* (Brenninkmeijer, 2010); a great part of Chap. 6 (*Neurofeedback as a Dance of Agency*) was published in *BioSocieties* (Brenninkmeijer, 2013); and part of Chap. 3 (*Glancing Behind the Scenes*, earlier published as Brainwaves and psyches) was also published in *History of the Human Sciences* (Brenninkmeijer, 2015). In all these 'peer review' processes, I received relevant comments, and I am grateful to the anonymous reviewers and editors of these journals. Moreover, I am also very grateful to the anonymous reviewers and the editors of Palgrave Macmillan who helped me to rewrite my manuscript and to turn it into a book.

References

Brenninkmeijer, J. (2010). Taking care of one's brain: How manipulating the brain changes people's selves. *History of the Human Sciences, 23*(1), 107–126. doi: 10.1177/0952695109352824.

Brenninkmeijer, J. (2013). Neurofeedback as a dance of agency. *BioSocieties, 8*(2), 144–163. doi: 10.1057/biosoc.2013.2.

Brenninkmeijer, J. (2015). Brainwaves and psyches: A genealogy of an extended self. *History of the Human Sciences, 28*(3), 115–133. doi:10.1177/0952695 114566644.

Contents

List of Figures

1

Introduction

'People, come on, start influencing your brain.' That idea is very marketable; it sells very well. In our culture and society it is a digestible concept. When people have a problem, they desperately want to recover and are prepared to try anything. And on the Internet you are bombarded with options (12).

This citation is a self-critical and reflexive phrase from a person (12) who tried multiple brain enhancers to solve his problems, especially concentration problems. He started with brainwave meditation, used several pharmaceuticals, tried various (brain) therapies, and bought some technical devices and computer games in order to manipulate his brain. Some of these devices had some effects, in the sense that they temporarily 'cleared his mind', and others had no result at all. Hence, this person is quite skeptical about certain brain devices and therapies, but this does not restrain him from trying other brain therapies. He is critical about some brain-enhancing practitioners and manufacturers, but he is also convinced of the need and possibility to change his brain.

One might wonder where this idea to change the brain comes from, and for what reason people keep on trying the recipe—even if they do not notice any immediate effect. The explanation of user 12 is clear: on the

J. Brenninkmeijer, *Neurotechnologies of the Self*,
DOI 10.1057/978-1-137-53386-9_1

Internet you are 'bombarded' with options. The message to take care of your brain is indeed widespread on the Internet and in other media. 'You think what you eat' warns a newspaper headline. 'Fool your brain' reads the heading of a news item about dieting. 'Reclaim your brain' orders an advertisement for brain games. 'Are your neurotransmitters out of balance?' asks a website for mood and energy management for women. 'Do you think this guy can be his real self with such a brain?' inquires the medical director of a brain clinic while he points at an awful-looking brain picture (SPECT) in an Internet commercial.[1] The same doctor contributes a more creative phrase: 'A lot of people think that mood problems are all in their head, that these are psychological problems. But they are not: they are brain problems.'

Spreading the message that people should work on their brains is not only a commercial trend, but also a serious quest of neuroscientific associations that, for example, organize 'brain awareness weeks' to make people more alert on their brains' capacities and how these can be to improved.[2] Such brain awareness campaigns followed a scientific development in which many problems that were formerly called 'mental' or 'social', such as depression, anxiety, hyperactivity, or learning disabilities, were reconceptualized as brain disorders (Conrad & Potter, 2000; Lane, 2006; Rose, 2007). The causes and solutions of these problems altered with the concepts: from life events to brain disturbance and from psychotherapy to brain interference (Clarke, Mamo, Fishman, Shim, & Fosket, 2003; Rose, 2007). These scientific developments, combined with a political drive to increase health and happiness in the population, contributed to increasing attention being paid to the brain.[3]

---

[1] SPECT stands for single-photon emission computerized tomography. These brightly colored brain images can be quite shocking since they present an image of the brain as having multiple holes and gaps if not in top condition. In another video the medical director phrases it like this: 'Do you think this guy can be his real self when the functioning in his brain looks like Swiss cheese?' (Amen Clinics Part 2 of 3). See also Johnson (2008) who discusses the work of this clinic, and uses the word 'moth-eaten' to describe the appearance of the images.

[2] The Society for Neuroscience and the DANA Alliance for Brain Initiatives, for example, organized brain awareness weeks, see www.sfn.org/index.aspx?pagename=baw_home and www.dana.org/brainweek/.

[3] The increasing attention devoted to the brain in scientific articles is easily demonstrated by typing the term 'brain' in Web of Knowledge (wokinfo.com), and performing a result analysis by year of publication. Besides the rise in publications over time, it is also striking that an almost threefold increase of publications can be observed from 1990 (6864 publications) to 1991 (19,445 publica-

These transformations were preceded by a shift in thought style regarding the brain. A few decades ago, the idea emerged that the brain was plastic and malleable, instead of stable and immutable as most neurologists thought before (Rubin, 2009). This change in perspective brought the promise of brain intervention as therapeutic intervention rapidly into action, since the brain was no longer seen as an organ that determined behavior, but also as an organ that could be trained or enhanced to change this behavior (Rubin, 2009). Moreover, the idea of a plastic brain transformed several academic disciplines because many researchers started to focus on the flexible neuro part of their work and formed new subdisciplines, such as neuropsychology, neuropedagogy, neuropolitics, or neuroeconomy, often subsumed under the name of 'neuroscience'. In other words, the shift in thought style in neurology from a brain that was static to a brain that was plastic gave rise to contemporary neuroscience (Abi-Rached & Rose, 2010; Pitts-Taylor, 2010).

As a consequence of these neuroscientific developments, people are increasingly taught that not only the cause of their behavior but also the solution for their problems is located in their brains, and hence they are more and more inclined to work on their brains to become better, happier, more peaceful, or smarter. Although user 12 might be more fanatic or experimental than others in his quest to improve his brain, he is not an exception. Many people nowadays try to manipulate their brains by using techniques varying from pharmaceuticals, to special diets or nutrition, games, or technical devices (e.g. Pitts-Taylor, 2010; Thornton, 2011). On the Internet, or in special 'brain clinics', several devices are obtainable that promise to improve people's brains and make them happier, healthier, or more successful. Light and sound machines, for instance, promise to switch someone's brain state into meditation, hallucination, or concentration (e.g. Heller, 1991). Neurofeedback is promoted to cure people from disorders like attention deficit hyperactivity disorder (ADHD) or to improve their performance in music or sports (e.g. S. Johnson, 2004; Mattin, 2006; Roberts, 2006). And devices that produce electric or magnetic current are

---

tions). The year 1990 was the start of 'the decade of the brain', as designated by former US president George H. W. Bush as the period in which the public awareness for brain research should be enhanced. This shows that the increased attention to the brain is not only a scientific and commercial development, but also a political trend.

said to aid people in recovering from their depressions, anxieties, physical pain, or sleeping problems (e.g. Harvey, 2004; Naish, 2007).[4]

Brains and Selves

Working on the brain to improve oneself brings an ontological difficulty. As the presented media quotes demonstrated, the brain seems to be turned into an entity that can be changed and fooled, whereas it can also be out of balance, disorder your mood, and keep you away from being your real self. That is, the brain has a passive and an active function. This raises the question of who or what responds to these actions of the brain. Listening to media quotes, neuroscientists, or brain device users immediately gives the answer: you, or the self, is the respondent of the brain. People work on their brains because they want to improve themselves. This, however, requires a distinction between the self and the brain, because the self tries to regulate the brain. While the self is reduced to the brain, it simultaneously is the operator of this brain. To state this more clearly: you have to take care of your brain, while your brain takes care of you.

I argue that this ontological distinction between one's self and one's brain should not be considered as a Cartesian distinction between something material (the brain) and something 'non-material' (the mind). But the monistic view, generally stated as 'the mind is what the brain does' (c.f. Churchland, 1986; Rose, 2007, pp. 192, 198), does not suit either because it veils the ambiguity between the brain and the self. In this book, I examine how people handle this ambiguity. I analyze how working on the brain operates on the self by studying people who seem to be convinced that their selves are (or are in) their brains, since they have decided to change themselves by manipulating their brains. As mentioned, people can swallow pills, take specific nutrition, do special brain games or trainings, or directly intervene into the working processes of their brains with technical devices. I am especially intrigued by those people who use technical brain

---

[4]To show that these claims reach a broad public, I referred to newspaper articles. However, there are also many Internet sites and scientific articles that make comparable claims (see, e.g., Arns, Ridder, Strehl, Breteler, & Coenen, 2009; Brunoni et al., 2012; Gruzelier, Egner, & Vernon, 2006; Nitsche & Paulus, 2011; O'Reardon et al., 2007; Ossebaard, 2000).

devices to improve themselves, since these techniques suggest direct inter-vention—without any bodily detours—in the working processes of the brain. As a consequence, using a brain device to improve oneself appears to create a very close connection between the brain and the self.

To examine how improving oneself by using a brain device operates on people's sense of self, I study therapeutic brain devices as contemporary examples of what Michel Foucault called 'technologies of the self': tech-niques people use to strive for their own health and happiness (Foucault, 1988). In his *History of Sexuality* (Foucault, 1990a, 1990b, 1992) Foucault described how since antiquity people had used techniques such as reading manuscripts, listening to teachers, doing confessions, or saying prayers to act on their selves and control their own thoughts and behaviors. Different techniques, Foucault argued, are based on different kinds of knowledge and ideas of self, and as a consequence, they also constitute different ways of being oneself (Foucault, 1988). To give an example, depending on the techniques people use—taking antidepressants, seeing a psychoanalyst, or confessing one's sins—they will see themselves as persons with chemical unbalances, repressed sexual desires, or struggles with the devil, which are three completely different ways of being oneself. Following Foucault, I wonder what kind of self is constituted by using a brain device.

Multi-sited Ethnography

To find out how users of brain devices think about, act on, and constitute themselves, I relied upon multiple sources and used multiple methods. That is, I made use of a multi-method qualitative study in which I did not study a subject, or a thing, but a subjectivity. As such, my methods can probably best be described as a 'multi-sited ethnography' (Hine, 2007, see also Marcus, 1995). This kind of ethnography differs from traditional ethnography because it combines several methods to study a social phenomenon, a subject, or a thing, which is not situated at one site (e.g. a tribe, a laboratory). The term is adopted by several scholars who not only describe, but sometimes also intervene in their research project, for example, by organizing workshops on the topic. Multi-sited ethnog-raphy is not meant as a method for 'objectively' or 'distantly' describing a phenomenon, but it shows the complexity of phenomena, including the influence the researcher can have on his or her studied phenomenon (Hine, 2007). Since the possibility to influence your research topic is quite an important issue when studying something as reflexive as the self, I reflect on this issue in the last chapter of this book.

This introduction will be followed by a chapter about brain devices that people can use to change themselves, namely, light and sound machines, non-invasive electric and magnetic stimulation, and neurofeedback. I give an overview of the uses, background, and scientific status of these technologies by studying how their effects were and are demonstrated. Hence, I give some insights in the problems expert practitioners have to attain scientific credibility and medical approval for their therapeutic claims. This chapter is mainly based on literature and Internet studies, and attempts to provide an overview and background of several 'neurotechnologies of the self'.

In the rest of the book, I focus on one of these devices, namely, neurofeedback—a computer system that makes people aware (with beeps and graphs) of their real-time brainwave activity so that they can try to change this activity.[5] To understand how this idea of working on oneself by working on one's brainwaves emerged, Chap. 3 explores the philosophies of four central figures in the history of neurofeedback. I studied the work and biographies of two principal scientists in the history of clinical electroencephalography (EEG)[6]: Hans Berger (1873–1941) who first visualized a human EEG, and William Grey Walter (1910–1977) who elaborated on Berger's work and showed that human brainwaves could be manipulated. Furthermore, I explored the experiments and media appearances of two psychologists who are seen as the 'founding fathers of neurofeedback', Joe Kamiya (1925–) and Barry Sterman (1935–). Next, I compared the ideas of these 'pioneers' of neurofeedback with the explanations of contemporary neurofeedback experts. This chapter is based on both ethnographic material and literature research. To find more information on the ideas

---

[5] The International Society for Neurofeedback and Research defines the therapy as 'NFT [neurofeedback training] uses monitoring devices to provide moment-to-moment information to an individual on the state of their physiological functioning. The characteristic that distinguishes NFT from other biofeedback is a focus on the central nervous system and the brain.' See http://www.isnr.net/neurofeedback-info/learn-more-about-neurofeedback.cfm (accessed on 16-7-2015) for the complete definition.

[6] Electroencephalography is the recording of the electroencephalogram, which is a visualization of fluctuations in electric brain activity. This electric brain activity is assumed to be evoked by neural activity.

of Walter than is available in Dutch libraries, I went to the archive of the Burden Neurological Institute (BNI) located in the Science Museum and to the Wellcome Trust Library, both in London. To understand the philosophies of Berger I collected his relevant publications and most (or all) of his published diary notes, and to understand the ideas and impact of Kamiya and Sterman, I read the relevant academic publications and searched the Internet and the database of LexisNexis to find information in newspapers and popular articles.

For Chap. 4, I mainly relied on user reports. By 'user' I refer to the person who is doing neurofeedback, who could be not only a neurofeedback client, but also a practitioner describing his or her own experiences with the therapy. I interviewed clients and practitioners in the Netherlands and the UK, and created an online questionnaire with open questions.[7] I questioned users about their introduction to neurofeedback, their reasons for doing or giving neurofeedback, and their ideas about how neurofeedback would solve their own or their clients' problems, and I asked them to describe their actions in the neurofeedback sessions. The interviews were taken at various places: at people's homes, at their neurofeedback clinics, by telephone, at the university, or at other places where the interviewees worked. Some interviewees not only talked about their personal experiences but also showed me how they tried to change their brain, and what had stimulated them: I saw neurofeedback games and privately taken EEG measurements; people explained their problems by showing me their brain maps, I saw books and articles that inspired people, and one user sent me computer files in which he kept notes and illustrations about changes in his feelings and brainwave activity, in different circumstances. To broaden my findings, I read hundreds of Dutch and English newspaper and magazine articles in which neurofeedback subjects were quoted, and I frequently checked websites of clinics and forums to collect reports and discussions of neurofeedback users.

To find out how practitioners work on the self of their clients, for example, by promoting and explaining their techniques to (potential) clients, I furthermore undertook several activities in the Netherlands and the UK. I visited open houses of neurofeedback clinics, initiated

---

[7] To protect the privacy of the users, I numbered all people cited and use these numbers as a reference after the quotation. See Appendix 1 for more information about the background of the users.

meetings in which practitioners explained their practices to students, attended a neurofeedback course for novice practitioners and other meetings for practitioners, observed a neurofeedback experiment performed to enhance the creativity of British school children, and underwent some neurofeedback sessions myself.[8]

The result of this multi-sited material is a book that presents a historical, ethnographic, and theoretical exploration of the mode of subjectivity that is constituted when people use neurofeedback to change themselves. To demonstrate that neurofeedback is not simply an alternative technique for eccentrics, but part of a society in which people increasingly start using techniques to manipulate their brains, I give a broader scope of brain devices in the forthcoming chapter. In Chap. 3, I explore how brainwaves and psyches (started to) interact in the work of early brainwave scientists and contemporary neurofeedback practitioners. I show how these people struggled with the connection between the self and the brain, introduced a machine-like version of the self, and combined materialistic philosophies with spiritual ideas. In Chap. 4, I focus on the ideas and acts of neurofeedback users to find out how working on the self with a brain device changes people's subjectivity. I argue that doing neurofeedback constitutes a new mode of being oneself, since the self is extended with a brain, and with various physiological, psychological, material, and sometimes spiritual entities that all start working upon the person's feelings, problems, and lives. Chap. 6 puts the constitution of this self in a broader context by focusing on the practitioners and by describing neurofeedback as a performance between human and non-human actors. In the final part, I reflect on my findings and draw

---

[8] To be precise about my (ethnographic) material: I attended two days of a four -day course for novice practitioners, in which seven participants and two supervisors were present (UK), I visited three open houses of neurofeedback clinics, one meeting for psychologists using neurofeedback, and five demonstrations of neurofeedback to students and other persons interested (Netherlands), and I underwent five neurofeedback therapy sessions myself (Netherlands). Furthermore, I made observations (one day) during a neurofeedback experiment that was performed on five schoolchildren (UK). I extensively interviewed one researcher, six practitioners (who were sometimes also researchers), and four clients, which altogether resulted in about 150 pages of transcriptions. More user reports were collected with an online open questionnaire which was answered by 13 clients. Most of the researchers/practitioners were also users, in the sense that they used neurofeedback to treat themselves (mainly for attention deficit disorder [ADD]) or in specific situations like before playing the guitar, taking an exam, or performing self-hypnosis. See also Appendix 1.

some conclusions. Overall, this book argues that working on the brain to improve the self does not reduce the self to the brain, but extends the self.

References

Abi-Rached, J. M., & Rose, N. (2010). The birth of the neuromolecular gaze. *History of the Human Sciences, 23*(1), 11–36. doi:10.1177/0952695109352407.

Arns, M., de Ridder, S., Strehl, U., Breteler, M., & Coenen, A. (2009). Efficacy of neurofeedback treatment in ADHD: The effects on inattention, impulsivity and hyperactivity: A meta-analysis. *Clinical EEG and Neuroscience, 40*(3), 180–189. doi:10.1177/155005940904000311.

Brunoni, A. R., Nitsche, M. A., Bolognini, N., Bikson, M., Wagner, T., Merabet, L., et al. (2012). Clinical research with transcranial direct current stimulation (tDCS): Challenges and future directions. *Brain Stimulation, 5*(3), 175–195. doi:10.1016/j.brs.2011.03.002.

Churchland, P. S. (1986). *Neurophilosophy: Toward a unified science of the mind-brain.* Cambridge, MA: MIT Press.

Clarke, A. E., Mamo, L., Fishman, J. R., Shim, J. K., & Fosket, J. R. (2003). Biomedicalization: Technoscientific transformations of health, illness, and US biomedicine. *American Sociological Review, 68*, 161–194.

Conrad, P., & Potter, D. (2000). From hyperactive children to ADHD adults: Observations on the expansion of medical categories. *Social Problems, 47*(4), 559–582.

Foucault, M. (1988). Technologies of the self. In L. M. Martin, H. Gutman, & P. H. Hutton (Eds.), *Technologies of the self. A seminar with Michel Foucault* (pp. 16–49). Amherst: The University of Massachusetts Press.

Foucault, M. (1990a). *The history of sexuality: The will to knowledge—Vol. 1.* London/New York: Penguin.

Foucault, M. (1990b). *The history of sexuality, vol. 3: The care of the self.* London/New York: Penguin.

Foucault, M. (1992). *The history of sexuality, vol. 2: The use of pleasure.* London/New York: Penguin.

Gruzelier, J., Egner, T., & Vernon, D. (2006). Validating the efficacy of neurofeedback for optimising performance. *Progress in Brain Research, 159*, 421–431.

Harvey, R. (2004). Mood disorders respond to new therapy. *The Toronto Star*, LIFE; p. C01.

Heller, Z. (1991). The electric kinesthetic innerquest; this is the ultimate head trip. *The Independent (London)*, December 15, 1991, Sunday, p. 2.

Hine, C. (2007). Multi-sited ethnography as a middle range methodology for contemporary STS. *Science, Technology and Human Values, 32*(6), 652–671. doi:10.1177/0162243907303598.

Johnson, D. (2008). "How do you know unless you look?": Brain imaging, biopower and practical neuroscience. *The Journal of Medical Humanities, 29*, 147–161.

Johnson, S. (2004). Wired up for mind games. *The Times (London),* 17 April, p. 16.

Lane, C. (2006). How shyness became an illness: A brief history of social phobia. *Common Knowledge, 12*(3), 388–409.

Marcus, G. E. (1995). Ethnography in/of the world-system—The emergence of multi-sited ethnography. *Annual Review of Anthropology, 24*(1995), 95–117. doi:10.1146/annurev.an.24.100195.000523.

Mattin, D. (2006). Good concept? It's a brainwave. *The Times (London),* Features; Body & Soul; p. 15.

Naish, J. (2007). Wake up -it's the instant-sleep machine. *The Times (London),* Features; Body & Soul; p. 3.

Nitsche, M. A., & Paulus, W. (2011). Transcranial direct current stimulation—Update 2011. *Restorative Neurology and Neuroscience, 29*(6), 463–492. doi:10.3233/RNN-2011-0618.

O'Reardon, J. P., Solvason, H. B., Janicak, P. G., Sampson, S., Isenberg, K. E., Nahas, Z., … Sackeim, H. A. (2007). Efficacy and safety of transcranial magnetic stimulation in the acute treatment of major depression: A multisite randomized controlled trial. *Biological Psychiatry, 62*(11), 1208–1216. doi: 10.1016/j.biopsych.2007.01.018.

Ossebaard, H. C. (2000). Stress reduction by technology? An experimental study into the effects of brainmachines on burnout and state anxiety. *Applied Psychophysiology and Biofeedback, 25*(2), 93–101. doi: 10.1023/A:100951 4824951.

Pitts-Taylor, V. (2010). The plastic brain: Neoliberalism and the neuronal self. *Health, 14*(6), 635–652. doi:10.1177/1363459309360796.

Roberts, G. (2006). Science & Technology: Free your mind; a neuroscientist claims he can unleash creativity by boosting low-frequency brainwaves—And he's tested the theory on 100 students at the Royal College of Music. *The Independent (London),* Features; p. 8.

Rose, N. (2007). *The politics of life itself: Biomedicine, power, and subjectivity in the twenty-first century.* Princeton, NJ: Princeton University Press.

Rubin, B. (2009). Changing brains: The emergence of the field of adult neurogenesis. *BioSocieties, 4*(4), 407–424. doi:10.1017/S1745855209990330.

Thornton, D. J. (2011). *Brain culture: Neuroscience and popular media.* New Brunswick, NJ: Rutgers University Press.

2

Brain Devices and the Marvel

Lord Lindsay got an enormous electro-magnet made, so large that the head of any person, wishing to try the experiment, could get well between the poles, in a region of excessively powerful magnetic force. What was the result of the experiment? If I were to say nothing! I would do it scant justice. The result was marvellous, and the marvel is that nothing was perceived. Your head, in a space through which a piece of copper falls as if through mud, perceives nothing. (Thomson, 1889, p. 261)

In the 19th century Sir W. Thomson—widely known as 'Lord Kelvin' and famous for his definition of the absolute zero temperature—called it a marvel that a subject whose brain was stimulated with an enormous electro-magnet perceived nothing. Today, however, such a result is one of the main frustrations of brain-stimulating scientists and practitioners. Although non-invasive brain devices are increasingly used in and outside of academic settings, expert practitioners have problems attaining scientific credibility for the therapeutic claims about their devices, and hence in getting them approved by medical agencies like the American Food and Drug Administration or the European Medicines Agency. In spite of the problems with scientific credibility and approval, these devices are easily accessible on the Internet or in brain clinics.

© The Editor(s) (if applicable) and The Author(s) 2016
J. Brenninkmeijer, *Neurotechnologies of the Self*,
DOI 10.1057/978-1-137-53386-9_2

People have many options to stimulate their brains with a device, without undergoing any surgery, and without seeing a doctor. They can, for example, try to change their brain frequencies with light and sound machines. They can also buy devices that work with electric or magnetic stimulation, or try to change their brainwaves by getting feedback on the working processes of their brain they are normally not aware of, with a neurofeedback device. These brain devices come in various sizes and prices, operate differently, and have diverse histories, but are more or less promoted for the same purposes. According to advertisements, websites, practitioners, and media articles, these devices help people to overcome psychiatric problems varying from depression, ADHD, burnout to anxiety, and they can enhance cognitive, artistic, or sports performances. Some of them come with spiritual promises, like evoking extrasensory perceptions or enhancing meditation skills, and they are sometimes also promoted as causing hallucinating or other 'mind-altering' effects. They are all presented as being completely safe to use, and almost without any side effects. The only side effect that is regularly mentioned is the possibility of evoking epileptic seizure in those who are vulnerable.

Light and sound machines are easily obtained on the Internet. They are also called mind or brain machines, or audiovisual entrainment, and consist of a pair of glasses with shifting LED lights and headphones with beeps that operate at a specific frequency intended to 'entrain' the brain's frequencies. People can buy or hire these machines, but sometimes they can also try them out at music and art events. Another option is to search on the Internet for instructions on how to build your own brain machine, for example, under headlines like 'hack your brain'. Because of their alleged stress-reducing and attention-increasing utilizations, light and sound machines are marketed for business trainings and excursions, but they are predominantly promoted for spiritual purposes (hypnosis, meditation), mental health problems, and cognitive enhancement.

Sometimes, brainwave machines are extended with low electric currents that you can put on your earlobes.[1] The idea behind this so-called

---

[1] See, for example, the 'DAVID PAL36 with CES' www.mindalive.com/2_1_8.htm (accessed on 13-11-2012).

cranial electrotherapy stimulation (CES) is that it sends electric current which is supposed to entrain the brainwave frequency on its way from one ear to the other. Instead of on the earlobes, you can also put the electrodes directly on the scalp, and use them as a transcranial direct current stimulation (tDCS) device, a technique that releases a low electric current on the cranium which is supposed to modify neuronal excitability and activity. tDCS is much more studied and tested than CES, but both techniques are promoted for psychiatric complaints like depression or anxiety, as well as for relaxation and cognitive enhancement. One can buy these brain technologies on the Internet or take the therapy in a clinic for the brain. However, the simplicity of the construction of these apparatuses did not escape the attention of its intended market. On forums, blogs, and videos on the Internet people discussed the best way to build a tDCS device for which you basically need only a sponge, a headband, a resistor, and a 9-volt battery.[2]

Some people also tried to build their own transcranial magnetic stimulation (TMS) devices.[3] These devices, which are mostly used in hospitals and academic labs, send electromagnetic pulses through the skull which are expected to create a flow of current that blocks or facilitates cortical processes. The American Food and Drug Administration approved one variant of this device—an apparatus that can give repetitive pulses (repetitive transcranial magnetic stimulation [rTMS])—as a therapy for 'adult patients with major depression who have previously tried medication and not improved satisfactorily' (Neuronetics, 2008). Because rTMS devices are quite expensive and are about the size of a dentist's chair only a few private clinics offer this therapy.[4] Nevertheless, some practitioners purchased the device and currently offer rTMS therapies to relieve people from their depression or post-traumatic stress disorder.[5] Furthermore, there are cheaper and more portable magnetic devices on the market, or in the making, like God (or Shiva) helmets that promise to

---

[2] See, for example, www.youtube.com/watch?v=I7nehK63Uk4 (accessed on 13-11-2012).

[3] See www.youtube.com/watch?v=HUW7dQ92yDU&feature=channel_video_title and www.youtube.com/watch?v=B_olmdAQx5s&feature=youtube_gdata_player (accessed on 13-11-2012).

[4] According to one practitioner prices vary from 20,000 to 70,000 euros.

[5] See for clinics in Canada and the Netherlands: www.mindcarecentres.com and www.brainclinics.com (accessed on 13-11-2012).

evoke religious experiences, migraine zappers that are supposed to relieve headaches, and thinking caps that are supposed to enhance cognitive performances.[6] Since tDCS appears to have the same effects as rTMS, some researchers are convinced that tDCS will become the smaller and cheaper variant of rTMS (Nitsche & Paulus, 2011).

Electric and magnetic brain stimulations are nowadays mostly used institutionally, and although the promise of self-help is tempting and people can buy or try to build these devices themselves, the majority of people who want to stimulate their brains prefer to do this without using electricity or magnetic current. For this group, neurofeedback can be a solution. Neurofeedback is a brainwave therapy in which the therapist (or the computer—see Chap. 6) 'reads' the brainwave activity of a client with an EEG (electroencephalograph), decides what brain parts[7] and frequencies to train, and adjusts the computer program in such a way that the client receives real-time positive or negative feedback on his or her brain activity with images or sounds. To give an example, a therapist might define that a client (e.g. diagnosed with ADHD) has too much low activity in (or in a certain part of) the brain, and asks the client to play a computer game (e.g. a racing game) that stops working whenever the brain (part) produces these 'wrong' slow frequency waves (Figs. 2.1 and 2.2). In this way, the brain is rewarded when it produces the right frequencies, and when it produces the wrong patterns, this reward stops. Instead of playing computer games, clients might also watch movies or listen to music (that stop(s) when the brain produces the wrong rhythm), watch graphics that, for example, turn from red (wrong) to green (good), or emoticons that can change from sad into happy. The therapy is currently offered predominantly by psychologists and is usually recommended for use by children with ADHD, but it is promoted for all ages and all kinds of psychiatric and medical disorders, as well as for cognitive enhancement and music or sport performances (e.g. Arns, Ridder, Strehl, Breteler, & Coenen, 2009; Coben, Linden, & Myers, 2010; Gruzelier, Egner, & Vernon, 2006).

---

[6] See, for example, www.shaktitechnology.com, www.healthcentral.com/migraine/treatment-256320-5.html (accessed on 13-11-2012) and (Macrae, 2008).

[7] That is, some therapists train the brain as a whole, other therapists work with (and try to balance) the left and right side of the brain, and again others work with brain areas.

All these brain devices—light and sound machines, cranial electric stimulation, tDCS, rTMS, and neurofeedback—have different histories and applications, but there are also some similarities. They are all proclaimed to be safe and easy to use and as able to solve psychiatric problems and suitable for self-enhancement. They inspire, and have inspired, many people to use them as a form of self-help. Moreover, they inspire people to not only help themselves, but also experiment on their selves, or their brains. Another characteristic, linked to this 'do it yourself' connotation, is that most brain devices are also used for spiritual and mind-altering purposes, such as improving meditation skills, or evoking hallucinations. Probably related to these spiritual, self-help, and self-experimenting practices is the circumstance that none of these therapies are completely accepted in the scientific or therapeutic world. In contrast to other treatments that are much debated concerning their efficacy, like psychotherapy or antidepressants (Greenwood, 1996; Ioannidis, 2008;

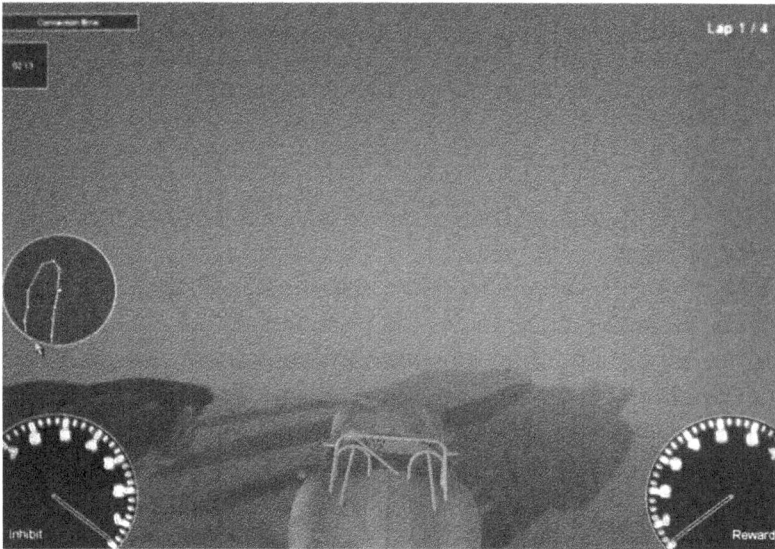

Fig. 2.1 Neurofeedback racing game. The client is connected to an EEG device that monitors the brainwave frequencies. If the 'right' frequencies are produced, the neurofeedback system makes the car speed up. If not, the car does not move, or is passed by another car (Used with permission of Roland Verment)

Fig. 2.2 Neurofeedback practitioner's screen. While the client plays a game or watches a movie, the practitioner keeps an eye on the client's brainwave frequencies. On the *left*, one can see two visualizations of the client's brainwaves; *top right* is the trend of the session, and in the four frames in the middle the requested frequency thresholds can be tuned. The bars represent the brainwave frequencies and go up and down. When they come above the threshold line a reward (or a withholding of the reward—depending on the training) follows in the form of a movement or sound (Used with permission of Roland Verment)

McGoey, 2010), brain technologies are not established as psychotherapeutic treatments.

To attain greater scientific approval, practitioners do their best to 'demonstrate' that their therapies are effective, for example, by performing experiments, by seeking collaborations with universities, and by presenting their techniques to a broad public on the Internet, or in other media. Magnetic stimulation (rTMS) and tDCS are increasingly gaining interest from universities and pharmaceutical companies, and neurofeedback is increasingly being subjected to clinical trials as well.[8] That is, there might be some problems concerning the scientific credibility of these devices,

---

[8] To compare, on July 7, 2014, Web of Knowledge gave 30,239 articles on transcranial magnetic stimulation (4420 when refined with research area psychiatry), 1637 on transcranial direct current stimulation, 994 on neurofeedback, 87 on cranial electrotherapy stimulation, and 22 on audiovisual entrainment.

but at least the latter three cannot be simply put aside as quackery. One could equally argue that they are still in their infancies.

A recurring question that pursues everyone using, selling, or (as in my case) studying brain technologies as therapeutic instruments is, 'Do they work?' The present chapter provides an answer to this question by focusing on the issue of scientific credibility. This will be done, not with an examination of the clinical effectiveness of these brain devices, or a presentation of the existing clinical literature, but with a historical analysis of how these devices were used and demonstrated. I rely on the concept of scientific demonstration (Ashmore, Brown, & Macmillan, 2005) to analyze the histories and contemporary uses of light and sound machines, electric and magnetic stimulation, and neurofeedback, and explore why these technologies have difficulties achieving scientific credibility.

Demonstration

According to Ashmore et al. (2005), practicing science is principally a form of demonstration: of making something visible. This can be done literally, for example, by showing something with a microscope or telescope. Another way of making something visible is with (brain) images like SPECT, PET, qEEG, or fMRI.[9] These images give the assumption that they directly 'show' something, while they are actually the results of very complex processes (see e.g. De Rijcke & Beaulieu, 2007; Dumit, 2004). Another indirect way to make something visible is by designing psychological experiments to imitate a 'natural' social situation to find out how human beings interact. Ashmore et al. discuss the 'problematic nature of demonstration within the psy disciplines' (2005, p. 78) and emphasize that the topic of such disciplines—basically the mind—is not directly observable, but also ubiquitous, which has the effect that everyone has some knowledge about it. They use the psychological controversy around false and recovered memories to discuss ways of demonstration and

---

[9] SPECT stands for single-photon emission computed tomography, PET means positron emission tomography, qEEG is quantitative electroencephalography, and fMRI refers to functional magnetic resonance imaging.

demarcation in the psychologies. The first mode of demonstration they indicate is this act of pointing out, of making something visible, which can be done by designing experiments, performing tests, or using a scan or microscope, but can also be accomplished by turning an idea or theory into an object of discussion or concern. This mode of demonstration mainly occurs in a private domain like a laboratory, or a therapy room.

In most cases, finding out something does not immediately lead to scientific or therapeutic approval. Scientists and other practitioners also have to reenact the discovery (Ashmore et al., 2005, p. 78). They have to make it appear again in front of an audience, for example, colleagues or the media, to gain witnesses for the phenomenon. In other words, they have to translate it into a different frame of reference (2005, p. 80; Stengers, 1997). This mode of demonstration can have some theatrical connotations because scientists have to exhibit that they exercise some control over the phenomenon. The location is no longer the private lab or therapy room, but is in the public domain.

The third mode of demonstration that Ashmore et al. have pointed out, is, what they call, a legal one (2005, p. 81). This form involves an advocate or a spokesperson who publicly aligns him- or herself to the phenomenon and who can act as the representative. That is, someone who speaks for the phenomenon. This way of demonstration sometimes literally takes place in the courtroom, but more often the phenomenon must be 'legally' adjudicated at other locations (the lab, the media). Ashmore et al. use the term 'legal' metaphorically. To legally demonstrate a phenomenon, it has to be defended against criticisms of, for example, colleagues, outsiders, or tests panels. To avoid misunderstandings concerning the metaphor of a 'legal' mode of demonstration, which is more useful to describe the 'memory wars', than for analyzing the scientific credibility of brain devices, I will call this mode 'polemic'.

Briefly stated, Ashmore et al. describe three modes of demonstration—showing, presenting, and speaking (or arguing) for phenomena—and the domains of these acts of demonstration are subsequently private, public, and polemic. They use this framework to analyze the 'memory wars' that were battled between two psychological disciplines: clinical psychologists who claimed that repressed memories (especially of sexual abuse) could exist and could be recovered in therapy and experimental psychologists who did not believe in recovered memories and claimed that creating false memories is not so complicated. In their article, Ashmore et al. give insights in scientific and psychological practices in general, but they also reveal some clear differences between experimental and clinical ways of demonstration.

The spokesperson for the false memories foundation is the experimental psychologist Elisabeth Loftus. Loftus designed experiments to demonstrate the possibility of false memories. In one of her well-known experiments, for example, she collaborated with family members of test subjects to create false memories about having been 'lost in the mall' as a child (Ashmore et al., 2005; Loftus & Pickrell, 1995). Loftus published her results in many books and articles and performed in television programs to reenact her findings to the public. Furthermore, she became a spokesperson in the courtroom to assist parents, or other relatives, who were accused of sexual abuse after their alleged victims had recovered their memories of the abuse in therapy. The three acts of demonstration are clear: Loftus showed something (false memories) in her experiments, staged her findings to the public, and became a legal spokesperson for her findings. With these forms of demonstration Loftus was very successful, and finally dominated the recovered memory debate.

The victory of Loftus in the memory wars is not that surprising. Ashmore et al. indicate that it is easier to achieve reliable witnesses for your findings in an experimental than in a therapeutic setting. In an experimental setting the experimenter decides which outcomes should be seen as 'interesting' and which as 'irrelevant', and this exclusiveness ensures that the ubiquity of the mind is circumvented. To give an example, everyone can say something about memory, but not everyone can recognize and demonstrate the outcome of a memory experiment. As an experimental psychologist, Loftus had many reliable witnesses in the form of experimental results, and she herself was also seen as a reliable representative, because she was supported by many colleagues, books, articles, and prizes.

It is more difficult to demonstrate what is created in a therapy room. Client statements, for example, are generally not considered very reliable, while the outcomes of an experiment are. Moreover, in a therapeutic setting there are only two witnesses: the client and the therapist. The social relations that are (officially) irrelevant in an experiment or laboratory are vital in the clinical setting. What is created in the therapy room is based on an interaction between the client and the therapist. As Ashmore et al. state: 'This creates a puzzle—how can what is demonstrated in the therapeutic setting then become amplified in and by the two other forms of demonstration (public and legal) when, by definition, the contingent social relation between therapist and client cannot be transported?' (2005, p. 96).

One strategy to transport findings (a recovered memory, for example) out of the therapy room is by treating them as case studies that confirm a particular theory. Individual case studies can be aggregated with other cases, which might upgrade 'the epistemic status to that of a statistic' (2005, p. 96). However, since these cumulated results can always be deconstructed to individual cases this also makes them vulnerable to criticism. Another strategy clinical therapists use to defend their therapies is emphasizing the personal character of clinical knowledge. Biographies, confessions, and feelings of clients are allowed to 'speak for themselves', because other people might recognize themselves in these narratives. Recovered memories, for example, were duplicated by people who had recognized themselves in the feelings and experiences of the personal stories of victims and organized themselves in solidarity groups. This strategy of creating a feeling of recognition is also used by experimental psychologists. Loftus' experiments, for example, were recognizable and replicable in the sense that people could try to create false memories—a 'lost in the mall' experience—in their relatives.

Emphasizing the individual, personal, recognizable, or replicable elements of a phenomenon might transfer this phenomenon from the private (the therapy room, the experimental setting) to the public domain. However, as I will demonstrate, such transformations are not always supportive of the scientific credibility of the findings. The present chapter uses the elements and domains of demonstration that Ashmore et al. have pointed out to analyze three kinds of brain devices. Ashmore et al. traced a genealogy of the experimental and therapeutic disciplines, and argued that the performances and later representations of Wundt and Mesmer exhibit the disputes between these disciplines—similarly, my analyses will also include contemporary and historical ways of demonstration. In my analyses, however, it appeared to be necessary to make a distinction between scientific and non-scientific realms of demonstration. One of the problems demonstrating the therapeutic effects of brain devices is that they are often well demonstrated in a private, personal, or self-experimental setting, but not always in a scientific realm.

Experimenting on the Self with Light and Sound

One way to utilize the three modes of demonstration that Ashmore et al. distinguish—showing, presenting, and 'speaking for'—is by constructing an impressive history for a particular field, phenomenon, or theory. Such histories start at very early ages to show that the phenomenon 'has always been there', are anecdotal so that people remember or recognize them and will 'reenact' the stories to other people, and refer to multiple important historical spokespersons that can speak for the phenomenon. Seen in this perspective, the effects of light and sound machines are well demonstrated. The histories of light and sound machines regularly start very early, for example, with prehistoric humans being hypnotized while dancing on the beat of drums in the light of flickering campfires. Another way to demonstrate (and reenact) the effects of light and sound is by referring to the recognizable hypnotic or dreamy feelings that can be caused by staring at flashing lights of disco balls (or campfires), and by listening to rhythmic drums, for example, in pop music. Furthermore, the history of light and sound machines is represented by several famous spokespersons.

One important pioneer in the development of light and sound machines is the British neurophysiologist and cyberneticist William Grey Walter, who demonstrated the effects of flickering lights on the brain. In his popular book *The Living Brain* (1953), Walter included a long and impressive history of the flicker effect, by suggesting that 'flicker'—according to him able to evoke epileptic fits—might have been crucial for human evolution:

> Oddly enough, it is not in the city but in the jungle conditions, sunlight shining through the forest, that we run the greatest risk of flicker-fits. Perhaps in this way, with their slowly swelling brains and their enhanced

liability to break-downs of this sort, our arboreal cousins, struck by the setting sun in the midst of a jungle caper, may have fallen from perch to plain,
sadder but wiser apes. (Walter, 1957, pp. 63–64)

Walter and colleagues had adopted the work of the German psychiatrist Hans Berger, known as the discoverer of the human electroencephalogram, and performed experiments in which they found that
these brainwaves could be increased by subjecting the brain to flickering
lights (Walter, 1957, p. 58). Gazing into a stroboscope, however, did
not only change the EEG, but also had unexpected side effects. In *The
Living Brain* (1953) Walter described how people who gaze with their
eyes closed into a stroboscope that flickers with an alpha frequency (8–12
Hertz) start to see visions, hallucinations, or 'waking dreams'. Depending
on the frequency of the stroboscope, Walter furthermore reported that
flicker could also evoke feelings like annoyance and anger, or cause epileptic fits, even in people who never had a fit before.

With *The Living Brain* Walter staged his findings to a broad public
and inspired many people in different ways. Neurologists were especially
interested in Walter's claim that flickering lights could evoke epileptic fits,
pedagogues discussed his work on brainwaves and feelings, and artists and
self-experimenters were intrigued by the possibility to cause visions and hallucinations with a lamp (Geiger, 2003; Hayward, 2001; Tanner & Inhelder,
1971). It was this third group of people—the artists and self-experimenters—that would give rise to the development of the light and sound machine.
Most of these people were interested not only in the hallucinogenic effects
of stroboscopic light, but also in the effects of mind-altering drugs.

The psychiatrist John R. Smythies, for example, tried and studied
the hallucinating effects of stroboscopic light, and suggested that flicker
could enhance the effect of mind-altering drugs (Smythies, 1959a,
1959b, 1960). Another famous reader of Walter's work was the author
Aldous Huxley, who wrote about his mescaline experiences in his books
The Doors of Perception and *Heaven and Hell* (1994, originally published in 1954 and 1956), and also mentioned the effects of stroboscopic light: 'To sit, with eyes closed, in front of a stroboscopic lamp is
a very curious and fascinating experience. No sooner is the lamp turned
on than the most brilliantly colored patterns make themselves visible'

(Huxley, 1994, p. 106). Inspired by Smythies, Huxley furthermore mentioned the fact that 'the stroboscope tends to enrich and intensify the visions induced by mescalin or lysergic acid' (Huxley, 1994, p. 106; see also Canales, 2011).

Walter, Smythies, and Huxley demonstrated the effects of light and sound in two domains. All three of them made the effect of stroboscopic light visible with experiments on themselves or test subjects, and successfully transferred these findings to the public domain with books and articles. Especially the books of Walter and Huxley reached a broad public and spread the word that flickering light could evoke drug-like effects. One reader of both authors was the writer William S. Burroughs who recognized the flicker phenomenon when he received a letter from his friend and artist Brion Gysin in which Gysin described a spontaneous hallucination. In 1958, Gysin had been traveling by bus through an avenue with trees when he had closed his eyes against the setting sun, which evoked a hallucination: 'An overwhelming flood of intensely bright patterns in supernatural colors exploded behind the eyelids (…) I was swept out of time' (Geiger, 2003, p. 11). Burroughs responded, 'We must storm the citadels of enlightenment' (Geiger, 2003, p. 11), and proposed to develop a machine that could procure the flicker effect. Together with the technician Ian Sommerville, Burroughs and Gysin constructed a 'dream machine', also called 'dreamachine'; a simple device made of a perforated cylinder turning around a bright lamp. According to Gysin, staring into the machine with eyes closed allowed people to see 'everything that can be seen, or has been seen, or will be seen' (Geiger, 2003, p. 54). He called it 'the very first exploration of one's own interior space' and since the dream machine was an art object that *makes* art, Gysin patented the device as a 'procedure and apparatuses for the production of artistic sensations' (Geiger, 2003, pp. 55, 66).

According to Gysin, the dream machine was a television broadcasting 'inner programming', and he hoped his machine would finally replace the TV (Geiger, 2003, p. 66). These commercial plans did not work out, but the machine has been exposed at several important art exhibitions.[10]

---

[10] After Geiger's book in 2003 and a documentary named Flicker in 2008, the traditional dream machine was released again in 2012. See www.dreamachine.ca/ (accessed on 13-11-2012).

The dream machine also influenced the use of stroboscopic light in the music scene and some art movies made use of its hallucinatory effect. One of these films was simply called *The Flicker* (1966), and was entirely based on flicker effects. It opened with warnings about the possible inducement of photogenic migraine and epilepsy. According to the director, the American artist Tony Conrad (1940), the movie was successful because of the unusual side effects the audience experienced: 'Some people saw birds. Letters or numbers. Many people saw concentric circles—the most common was colored, jiggling mandala figures.' Other effects reported besides hallucinations were 'phenomena of addiction' and people becoming 'uncannily frozen' (Geiger, 2003, p. 75).

Burroughs, Gysin, and Conrad can be seen as spokespersons for flicker or the dream machine who demonstrated the phenomenon to a broad public. However, they especially accentuated the personal experiences, and, plausibly, this has hindered flicker attaining scientific credibility. All three spokespersons were famous, but they were also associated with drug experiments and experiences. Moreover, these representatives promoted the stroboscope itself as a device to evoke drug-like experiences. Flicker effects were made visible in several private domains, like labs, living rooms, or ateliers. They were reenacted in public domains like galleries and cinemas, and promoted and perhaps defended by famous representatives. However, flicker devices were especially demonstrated as producing personal mind-altering or artistic effects. A device that is promoted to produce one's 'own interior space' does not create reliable witnesses since people have no experiences to share or to recognize. Hence, people like Gysin, Burroughs, and Huxley actually transported the flicker phenomenon out of the scientific and into the personal realm.

This development from the public and scientific to the personal domain continued. Not long after flicker made its appearance in art installations which allowed the audience to collectively stare into a stroboscope and have their own experiences, the first personal machines were produced. One of the first devices that combined flicker with sound pulses was the Synchro-Energizer, a device with several goggles and headphones so that people could use them privately. This device that was constructed in the seventies and patented in the early eighties, and was promoted for creativity, meditation, and relaxation. Shortly thereafter, other portable devices entered the market with names like 'Relaxman', and DAVID 1 (Digital Audio-Visual Integration Device) (Hutchison, n.d.). These so-called brain or mind machines became popular at disco and electronic dance parties. In the media they became known as 'digital drugs' because the machines were supposed to provoke a hallucinatory effect, comparable with the effect of LSD (Geiger, 2003).

Through the years, light and sound machines became promoted for more and more purposes, and at the moment they are advertised for almost everything: to enhance academic and sport performance, to improve hypnosis and meditation, or to reduce symptoms of 'Stress and Anxiety, Post Traumatic Stress Disorder (PTSD), Attention Deficit Disorder (ADD), Pre-Menstrual Syndrome (PMS), Chronic Fatigue Syndrome (CFS), Seasonal Affective Disorder (SAD), Depression, Insomnia, Autism, Chronic Pain and Fibromyalgia'.[11] The portable sizes of light and sound machines, and the relatively low cost, allow one to use these devices where and whenever one wants. Moreover, people can also build these devices themselves, or simply bypass the whole machine and download some software to watch the flashes and hear the beeps from their PC.[12] Although it is hard to find out how many persons nowadays use light and sound to entrain or 'hack' their brains, for example, in the hope of enhancing their memory, improving their meditational skills, or suppressing their fears, it is obvious that these so-called 'brain machines' allow people to experiment or work on their brain without needing to see a practitioner, therapist, or teacher.

The history of the light and sound machine is a history of self-experimentation. Not only did artists, writers, and scientists use the stroboscope to get into a certain state, they inspired a wide public to experiment on themselves, for example, by staring into the dream machine, watching flicker movies, or experiencing hallucinating effects at music parties. Some early promoters of the light and sound machines—Burroughs, Huxley, Smithies—are also important figures in the history of mind-altering drugs, and until today this connection between light

---

[11] Retrieved from www.mindalive.com, in December 2011.

[12] Nowadays, people can also use the Internet to evoke hallucinatory experiences, or upload mp3s with names like 'marihuana', 'cocaine', or 'ecstasy' (see, for example, www.i-doser.com, accessed on 13-11-2012).

and sound effects and drug experiences exists, for example, in newspaper articles that publish about teenagers getting a 'digital high' by using light and sound (e.g. Hesse, 2010). Even today light and sound machines are especially promoted and used for self-help and self-experimentation, and as a consequence—because they were especially promoted in the personal, and not in the public scientific domain—these devices never received much scientific attention.

Electric and Magnetic Demonstrations

In contrast to light and sound machines, which were especially demonstrated in the personal domain, expert practitioners of electric and magnetic devices emphatically try to demarcate and demonstrate their technologies in a scientific realm. Nevertheless, self-experimenting practices can also be traced in the history and contemporary uses of electric and magnetic stimulation. At the end of the 19th century, something like a 'self-help market' for electric and magnetic tools existed (Loeb, 1999), and nowadays some people share their experiences about home-made electric or magnetic brain-stimulating devices on the Internet. However, this is not the information that you will find when reading handbooks or articles on electric and magnetic stimulation. On the contrary, submerging yourself in the world of electric and magnetic devices will convince you 'not to try this at home', since an expert is needed to perform the miraculous working of these techniques.

Just as in the case of the light and sound machine, presented histories of electric and magnetic stimulation generally start early. In handbooks and articles on the therapeutic promises of electric brain stimulation, timelines sometimes start just after Christ with a Roman court physician using electric torpedo fish to treat patients suffering from headaches and gout (e.g. Pascual-Leone & Wagner, 2007). Or they begin in the 18th century when Luigi Galvani performed experiments with an electrochemical cell to stimulate animal tissue (e.g. Brunoni et al., 2012; Horvath, Perez, Forrow, Fregni, & Pascual-Leone, 2011). Some anecdotal experiments and discoveries in the 19th century are repeatedly mentioned, like Aldini's electric therapy and his attempts to waken the dead (1804),

Duchenne de Boulogne's experiments to stimulate facial muscle movement with electrodes (1862), and Bartholow's success to stimulate an exposed human cortex with a small electric current by which he produced muscle movement in the patient's body, and caused a fatal coma (1874) (e.g. Brunoni et al., 2012; Higgins & George, 2009; Horvath et al., 2011; Pascual-Leone & Wagner, 2007). From there, the timeline regularly takes a leap to the introduction of electroconvulsive therapy in 1937, and to Delgado's stimulation of animal brains that he demonstrated by stepping into the ring with a remotely controlled bull in 1965. From the sixties, a period in which a 'neuromolecular gaze' emerged which gave an enormous impulse to brain research and formed the neurosciences (Abi-Rached & Rose, 2010), many different experiments are mentioned (e.g. Brunoni et al., 2012; Higgins & George, 2009; Horvath et al., 2011; Pascual-Leone & Wagner, 2007).

The history of magnetic stimulation is regularly connected to the history of electric stimulation with just a few magnetic experiments mentioned, such as Faraday's demonstration of electromagnetic induction (1831), D'Arsonval's findings that magnetic stimulation of the human head could produce phosphene, vertigo, and syncope in the subject (1896), and Thompson's findings that magnetic stimulation of the human brain could evoke flicker perceptions (1910). From there, the history generally jumps to the experiments of Anthony Barker who developed the first magnetic pulsing device in 1976, and the first TMS device in 1985 (Brunoni et al., 2012; Higgins & George, 2009; Horvath et al., 2011; Pascual-Leone & Wagner, 2007).

In these histories of electric and magnetic stimulation several acts of demonstration can be traced. In all experiments something was made visible—privately to the subject (flicker perceptions) or as could be observed by an audience (muscle movements). Some of these experiments, moreover, were reenactments and had some theatrical connotations (Ashmore et al., 2005), like the stimulations that occurred in the presence of an audience, or that almost turned into a circus act, such as Aldini's electric stimulation of dead bodies, or Delgado's taming of the bull.

The third mode of demonstration, the act of representing or speaking for a phenomenon, can also be observed in these histories. The whole timeline is actually an enumeration of representative cases that can speak for the phenomenon, while excluding others. One important promoter of magnetic stimulation who is left out by TMS promoters and only mentioned in relation to TMS by skeptics (e.g. Szasz, 2006), historians (Vijselaar, 2007), or unwanted representatives like 'magnetotherapists' (Rosch, 2009) is Franz Anton Mesmer. This 18th-century German physician used magnets and hand movements to cure his patients from hysteria and other maladies. Mesmer's practices are frequently maligned and ranged in the realm of pseudoscience, but it is likely that they have encouraged the idea of using magnets for treating people.

Other practices that were probably influenced by Mesmer, and that are not mentioned by contemporary electric and magnetic stimulators, are the electric and magnetic tools sold at the end of the 19th and the beginning of the 20th century, in Europe and the USA. The idea of electricity and magnetic power that gives energy to the body and the mind was very common in this period, and resulted in a self-help market of tools like electric flesh brushes, electropathic belts, and magneto-electric batteries (e.g. Peña, 2001). Reading advertisements demonstrates that these devices were mostly promoted for physical problems like gout, impure blood, and kidney problems, but also for problems that most people would nowadays call 'mental'. In one advertisement, people with 'sedentary habits and weakened nervous powers' are addressed, and another ad declares that 'all who suffer hysteria or any form of muscular or organic nervous weakness should stop taking poisoning drugs and quack medicines and try the healing, strengthening, exhilarating effect of mild continuous currents of Electricity'[13] (Figs. 2.3 and 2.4).

The official as well as the unofficial history of electric and magnetic stimulation has some theatrical elements. Delgado and Aldini were not the only showoffs, in the sense that they created a spectacle with their experiments; Mesmer perfectly fits this picture since he attached his patients to a magnetic vat and tried to cure the fainting and screaming people by slowly dancing amongst them (Ashmore et al., 2005). Moreover, contemporary promoters of electric and magnetic stimulation sometimes also perform theatrical acts to demonstrate their devices to the public.

---

[13]Advertisements were found in the Wellcome Trust Library, in London. See, for example, http://catalogue.wellcomelibrary.org/record=b2018563-S12

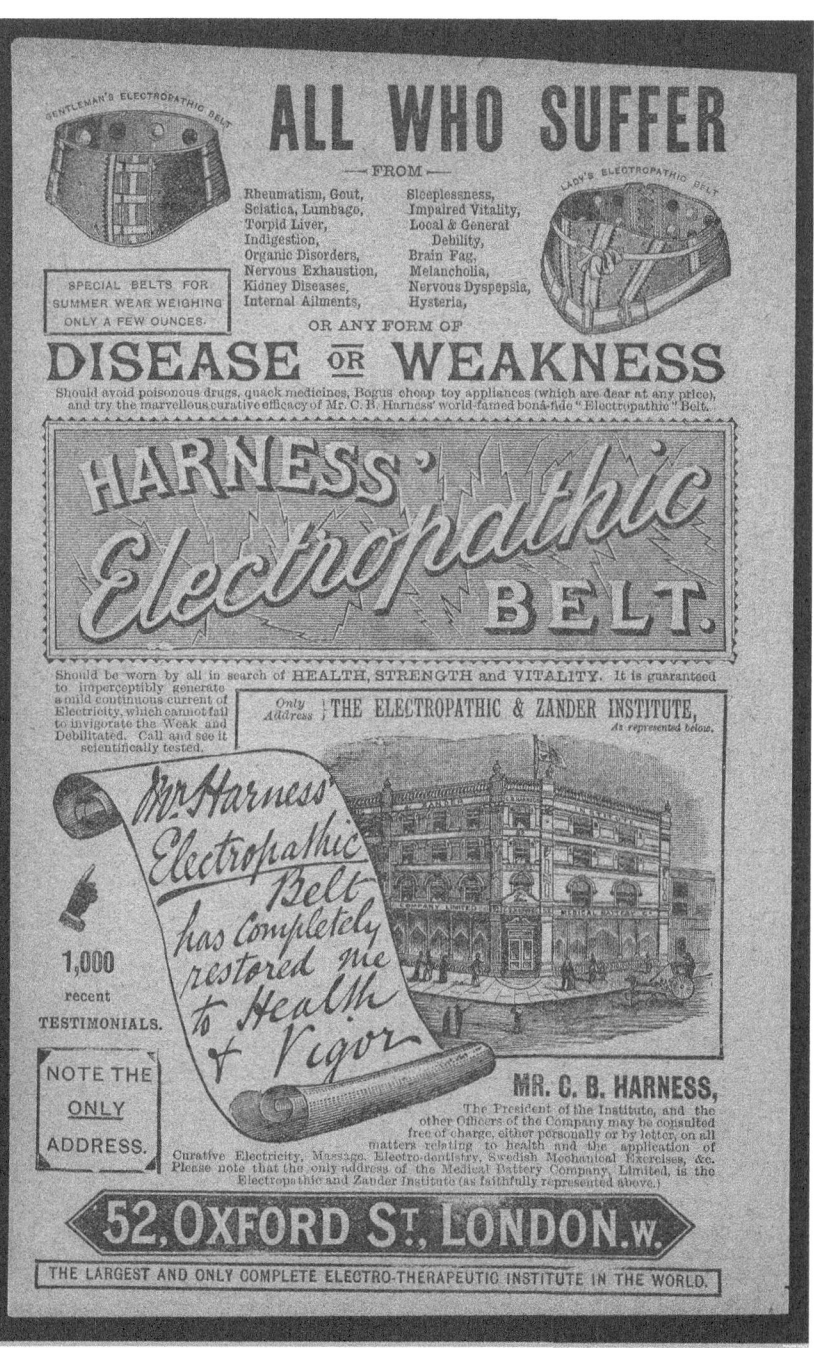

Fig. 2.3   Advertisement for an electropathic belt (Source: Wellcome Library, reproduced under a CC-BY 4.0 License)

Fig. 2.4 Advertisement for an electropathic belt (Source: Wellcome Library, reproduced under a CC-BY 4.0 License)

TMS in particular is an expensive technique that is often explained in somewhat grandiloquent language. During a TMS workshop for psychological researchers that I attended, for example, the demonstrator explained that 'no one knows what TMS does', except that 'we are messing up cortical processes in the brain'. He explained that using the repetitive version of TMS as a research tool is not possible 'because of the risks', and warned that, since epileptic seizures are possible, researchers using TMS need a first aid certificate and a good insurance before they start experimenting. During a course for psychology students this researcher took about 30 minutes to inform his audience about the risks of TMS before he tested his device on the brain of one of the volunteers. The only thing that this demonstration seemed to produce was a short 'tick' of the machine, and the volunteer reported not having felt anything.

This case, in which the demonstrator made much fuss of a stimulation that ostensibly resulted in nothing, is perhaps best understood by returning to Lord Kelvin, quoted at the start of this chapter, who called it a marvel that nothing was perceived after the subject's brain was stimulated with an enormous electro-magnet. Likewise, the TMS demonstrator suggested that it is a marvel that TMS messes up cortical processes in the brain while subjects perceive nothing. During an open day of a clinic offering rTMS for depressed patients that I attended in 2009, a comparable 'nothing' message was presented. Not by stressing the risks of TMS this time, but by focusing on its safety. A psychologist asked a volunteer to sit down in the rTMS chair and calmed him down with 'relax and have a nice look out of the window'. She explained to the audience that getting rTMS is often described as the sensation that a dwarf is jackhammering into your brain, and thereupon calmly demonstrated that she could move the thumb of the man by stimulating his head. Again, something marvelous was demonstrated: it is a marvel that rTMS can cure people's depressions and move people's thumbs while it 'only' feels like a dwarf jackhammering into the brain.

Without the information of these promoters, the audience could easily think that TMS really does nothing, since the subject does not show any reaction, or people can think that it really is dangerous since it moves the subject's thumb. Hence, these promoters are essential to demonstrate what TMS is and does. They are needed to teach others what is interesting in their experiments, and why it has succeeded. In a similar way, representatives explicate the histories of their devices and define that the healing practices of Mesmer or electropathic belts do not belong to their histories, while torpedo fishes and bull fighting do.

However, researchers and expert practitioners are not the only people who do their best to demonstrate the effects of tDCS and rTMS devices. On YouTube several hobbyists demonstrate their home-made electric and magnetic brain devices by stimulating their head with a magnet or battery—often with an ostensible 'nothing happens' result—and on other Internet forums users inform each other about how to build these devices, and where to put the electrode for what purposes. One might argue that these movies show that electric and magnetic devices are very well demonstrated in the public domain, since people adopt the theories of TMS and tDCS researchers and replicate their experiments. However, it can also be argued that these hobbyists do not simply replicate experiments, but actually demonstrate them to an audience not intended by the official promoters, which transfers these devices into another—personal and non-scientific—realm. The fact that scientists and expert practitioners are not pleased with the demonstrations of lay people but warn against them (e.g. Brunoni et al., 2012) suggests that they indeed try to keep these devices out of the personal domain.[14] Apparently, expert practitioners try to demarcate electric and magnetic brain stimulation as a professional practice; something not to be tried out at home.

Expert practitioners use several strategies to demonstrate electric and magnetic brain therapies to the public domain. They carefully construct the histories of their devices, stage their therapies to the public, and try to keep them out of the personal, self-experimental domain. However, continuing controversies concerning the efficacy, safety, research methods, and best applications and localizations make these techniques difficult to represent in the polemic domain.[15] Although some effects are clearly demonstrated and speak for themselves—moving the thumb, creating

---

[14] To give another example, www.mindalive.com sells a device for Audio-Visual Entrainment (AVE), CES, and tDCS but warns: 'CAUTIONS: tDCS is very powerful and if applied improperly, can result in negative side effects. Therefore, the sessions for tDCS will only be released to qualified clinicians' (accessed on 13-11-2012).

[15] To give an example, the National Health Council of the Netherlands evaluated rTMS as an effective treatment for depression in 2008, while the Health Care Insurance Board decided in 2011 that insufficient data existed to state that rTMS is an effective study for depression.

flicker perceptions—other effects such as curing depressions, solving anxieties, or improving memories still need to be justified.

Neurofeedback as a Spiritual Science

Different from light and sound machines and electric and magnetic stimulation, neurofeedback is not demonstrated with an age-old history that forms an exposé of famous representatives and anecdotal experiments. On the contrary, reading Dutch newspaper articles on neurofeedback gives the impression that a brand new brain therapy is establishing in the Netherlands. Doing the same with English-language newspapers, however, evokes the suggestion that neurofeedback is old wine in a new bottle, and probably not a very good one either. Perhaps both impressions are right. Before 1997, studies of neurofeedback (or EEG biofeedback, as it was called) almost exclusively derive from the USA and Canada. European countries are making up arrears since then, and the Dutch have started to contribute in the last few years.[16] Until today, neurofeedback is mostly practiced in the USA, but the density of neurofeedback clinics has become highest in the Netherlands. In other words, neurofeedback is a relatively new therapy in the Netherlands, but not in the USA.

Perhaps this difference also clarifies why American newspapers appear to be more critical than the Dutch, for instance, by discussing the costs in terms of money and time, and the lack of regulation (e.g. Ellison, 2010). Although not presented in this way to the general Dutch public, in the Netherlands many people also agree that neurofeedback is an expensive and time-consuming therapy that lacks regulation. Costs easily run up to 3000 euros for a treatment and these costs are usually not (completely) covered by insurance companies.[17] Clients are expected to

---

[16] Analysis made with Web of Knowledge.

[17] Costs are variable. In the Netherlands, one session costs around 65–100 euros. Some persons need 20 sessions, others 70; in general they take 30–40 sessions. In most clinics clients also get a qEEG which costs around 500 euros. People are only covered for these costs if they receive their neurofeedback therapies from registered psychotherapists. Otherwise, some reimbursement is possible if the neurofeedback is called 'coaching' or 'alternative therapy'. In practice, Dutch clients pay about half of the costs themselves. In the USA, insurance companies generally do not cover neurofeedback (Information retrieved from interviews with Dutch practitioners).

do their 1-hour trainings once or twice a week, and this often for 30–40 sessions. The lack of regulation is an annoyance among practitioners who, for example, blame each other for not using the right programs or method, or for offering the therapy for complaints that have never been tested with neurofeedback. That is, while some practitioners stress that neurofeedback has proven successful only for ADHD, epilepsy, and sleeping disorders, many others offer the therapy for almost everything: mental and physical problems, spiritual purposes, and even for peak performances in sports or business.

Such disagreements about methods, results, and definitions of neurofeedback are most likely related to the lack of a standard neurofeedback certification. Although many neurofeedback practitioners have a background in psychology, this is not a requirement. Moreover, everyone who is interested in neurofeedback can do an introduction course (or not) and start a neurofeedback clinic. There are many of such courses offered in Europe and the USA, and although some certifications are better credited than others, there are no strict standards.[18] For this book, I attended a four-day neurofeedback course for beginners in which the participants came from very different backgrounds.[19] Some of them were psychologists or therapists, but there was also a mother of a child diagnosed with ADHD, an accountant, a teacher, and a mental coach for sportsmen. When discussing the equipment, the course supervisors recommended some neurofeedback systems varying in price from 1400 to 3000 pounds, and gave the advise to 'Buy a nice cheap machine, train yourself and your friends or family members, [and] when you are used to it and want to start a practice you buy a more expensive machine'. In other words, in theory it is possible that a neurofeedback client is treated for complaints that have never been tested with neurofeedback, by an accountant, housewife, or teacher who has followed a four-day course,

---

[18] For example, in the Netherlands, the section neurofeedback of the Dutch Psychological Association uses the criteria of the American 'Biofeedback Certification International Alliance'— an online exam for biofeedback specialists—for their register. However, at the time of writing, the association had only 30 therapists registered while many more therapists offer the therapy (See http://www.psynip.nl/sectoren-en-secties/intersector/neurofeedback.html and http://www.bcia.org/i4a/pages/index.cfm?pageid=1).

[19] That is, I attended two days of this four-day course.

invested 1400 pounds, and learned the ins and outs of neurofeedback by self-experimentation.

Probably related to this lack of regulation is the lack of scientific agreement, or perhaps even scientific proof for this therapy. In spite of all claimed therapeutic results, the increase in the number of clinics and clients, the growing number of articles and scientific associations, and the efforts of researchers to make this therapy 'evidence-based' (Arns et al., 2009), the scientific results of neurofeedback are still under debate (e.g. As van, Hummelen, & Buitelaar, 2010; Huitema & Eling, 2009; Logemann, Lansbergen, Van Os, Böcker, & Kenemans, 2010; Loo & Barkley, 2005; Vollenbregt, van Dongen-Boomsma, Buitelaar, & Slaats-Willemse, 2014)

In interviews with expert practitioners, various explanations are brought up to explain why scientists have such problems demonstrating the effects of neurofeedback. Some practitioners state that in contrast to competing devices like psychopharmaceuticals, neurofeedback is not supported by the pharmaceutical industry which makes it more complicated to attract the same attention. Others refer to the difficulty of making neurofeedback placebos to develop the control trials that are needed to make their therapy evidence-based. Another argument is the lack of regulation which allows everyone to start a clinic without the need of any degree or training. One researcher specifies that especially those practitioners trained in the USA use old-fashioned methods and programs, while practitioners trained in the Netherlands or the UK use better developed software and often take a quantitative EEG (qEEG)[20] before they decide what to train.

Differences between the USA and Europe in the demonstration of neurofeedback are probably related to the history of this therapy. In the seventies and eighties, EEG biofeedback was one among several biofeedback techniques that were quite popular, especially in the USA. Watching your fluctuating brainwaves with the purpose of getting them under control was no more special than other biofeedback techniques which gave people information about their blood pressure, respiration, skin temperature, or heart pulses, in order to reduce these. People mainly tried EEG

---

[20] A qEEG is an EEG that is (automatically) analyzed and compared with a standard and visualized in an understandable image. Instead of incomprehensible brainwaves, it shows heads with green (normal), red/yellow (high), or blue (low) activity.

biofeedback to enhance their alpha waves, which was understood as a state of peace, relaxation, or meditation. In the nineties, also called the decade of the brain, the attention to the neuro part of biofeedback started to grow, and this trend continues until today.[21]

Unlike the alpha trainings of the seventies, nowadays people strive for a so-called good and healthy brain in which frequency amplitudes—alpha, beta, theta, delta, and sometimes gamma[22]—are compared and normalized. The main focus is no longer on relaxation and meditation, but on treatment and self-enhancement. Still, however, neurofeedback is often compared with meditation and users regularly also practice, or are interested in, alternative techniques like hypnosis, acupuncture, yoga, or meditation. Reading newspaper articles on neurofeedback gives the impression that the spiritual connotation is stronger in the USA than in the Netherlands. That is, American newspapers use more terms like 'meditation', 'yoga', 'acupuncture', 'telepathy' in articles on neurofeedback. It is conceivable that this (stronger) association has hindered the credibility for neurofeedback in the USA, which resulted in a dissimilar technical and scientific development in both areas.

This spiritual connotation is rooted in the first EEG biofeedback experiments. The American researchers Elmer and Alice Green, for example, went to India to study the physiology of yogis, fakirs, and sadhus (Green & Green, 1978), and the American psychologist Joe Kamiya studied the EEGs of masters in Zen and meditation (Kamiya, 1968). These studies were not only demonstrations in the sense that they made something

---

[21] A Web of Knowledge analysis demonstrates that the term 'neurofeedback' was seldom used in the 1970s, made its appearance in the 1980s, rose in the 1990s, and its use has rapidly increased during the last decade, also called the 'decade after the decade of the brain' (See, for example, www.dana.org/news/cerebrum/detail.aspx?id=25802 accessed on 15-11-2012).

[22] According to contemporary neurofeedback experts, training different brain frequencies can produce different mental states. Normally, alpha waves have a frequency range from 8 to 12 Hertz and when these dominate it gives a feeling of peacefulness. Increasing the amplitude of your beta waves (13–21 Hertz) makes you more focused, high beta (20–32/38 Hertz) leads to hyperalertness, theta waves (4–8 Hertz) increase your creativity, delta waves (1–4 Hertz) normally occur mainly during sleep, and gamma waves (38–42 Hertz) are said to correspond with cognitive processing. Problems occur when these waves are not in balance anymore. A brain that shows high beta waves when the subject is asked to relax can reveal that the person is stressed or anxious. Too high alpha and theta can refer to attention deficit (hyperactive) disorder (ADD/ADHD) or depression, and delta waves during waking hours can indicate brain injury (Demos, 2005). If one of these is the case, neurofeedback can be the solution to bring these frequencies back to normal.

visible—that is, spiritual figures controlling their brainwaves—but they were also reenacted in the public domain. The experiments of Elmer and Alice Green were shown in the movie *Biofeedback: The Yoga of the West* (1974), and Kamiya's experiments were published in a popular magazine. These forms of enactment to the public were quite successful since they encouraged many people to try to enhance their own brainwaves. Besides these spiritual investigations, however, there are also more 'down to earth' studies of American pioneers who, for example, trained the brainwaves of cats (Wyrwicka & Sterman, 1968), hyperactive children (Lubar & Shouse, 1976), and epileptic patients (Sterman & Macdonald, 1978). These studies are nowadays reenacted in books and articles that present neurofeedback histories. That is, the spiritual and the 'scientific' experiments are both well demonstrated to the public.

Scientific and spiritual aspects can still be experienced when visiting a (Dutch) neurofeedback clinic. It is almost impossible to do neurofeedback without being confronted with spirituality, for example, in the form of music during the training, conversations with the practitioner, or magazines in the waiting room. On the other hand, it is equally difficult to do neurofeedback without being confronted with the scientific background and proclaimed evidence for neurofeedback, for example, in the form of studies on neurofeedback websites, books in the waiting room, or conversations with practitioners. This outwardly smooth connection between science and spirituality, brains and spirits, materialism and mentality is very characteristic of neurofeedback.

Neurofeedback practitioners do not only allow spiritual accents in their clinics, they also permit certain forms of self-help and experimentation. Users mostly go to a clinic to let their brains be trained by a practitioner, but they can also do the therapy at home via tele-neurofeedback, buy the equipment to train themselves, or start their own clinics and become practitioners themselves. It can be argued that neurofeedback experts are not very good at demarcating their therapies as scientific (and not spiritual) and professional (and not experimental) practices. As in the case of light and sound machines, neurofeedback appears to be well promoted in the personal (self-help, spiritual) domain, and in contrast to electric and magnetic devices, this therapy is not carefully demarcated, and hence not so well demonstrated in the scientific domain.

This problem of demarcation might be related to the situation that neurofeedback is a collective performance of a practitioner and a client. Moreover, to make the therapy a success, a certain form of self-help—in the sense of the cooperation of the client—is required. In contrast to users of light and sound machines or electric or magnetic stimulations who passively (or let others) stimulate their brain, the neurofeedback subject has to be an active subject who trains his or her own brain. Hence, to demonstrate that neurofeedback works, practitioners are partly dependent on the acts of their subjects, and since self-help and spirituality can help the user, these practices are encouraged, instead of deterred.

Ashmore et al. discussed the problem therapists have in demonstrating their therapies outside of the therapy room. In the case of neurofeedback too this problem emerged. The success of the demonstration depends on the performance of the subject, and is partly a result of the relation between the therapist and the client. For example, practitioners instruct and motivate their clients and help them understand what has changed in their brains and lives. This therapeutic performance probably benefits from spiritual and experimental elements, since these can help to motivate, instruct, or relax the client. However, to demonstrate that neurofeedback is an 'evidence-based therapy'—which would increase the scientific credibility, and hence the chance of being covered by insurance companies—expert practitioners have to develop experimental settings in which the client–therapist relation is subordinated. That is, the scientific (and financial) quest and the therapeutic quest seem to hinder each other.

To demonstrate the effects of their therapies neurofeedback practitioners have to effect changes in the (private) therapy room, and reenact them in the public domain. They stage their results, for example, by organizing open houses for their clinics, putting information on the Internet, and giving media performances. During these presentations, they often refer to case studies and personal experiences of clients, and hence they emphasize the individual and personal character of their therapies. This strategy is useful to convince potential clients who can recognize themselves in the problems of other neurofeedback clients, but does not contribute to the scientific credibility of these devices. To attain scientific approval and insurance coverage—at least in the Netherlands—practitioners have to make their therapies 'evidence-based'. Hence, some expert practitioners try to reenact their therapeutic results in experimental settings. They do this by offering their assistance (and technologies) to universities, and starting their own (PhD) studies, and they present these results to a public scientific domain by publishing books and articles, and organizing conferences. In spite of all efforts, and all claimed clinical successes, neurofeedback is not a therapy with high scientific credibility. What is lacking is perhaps the representative spokespersons—aversive to spiritual and self-help practices—who defend these therapies in the polemic scientific domain.

Brain Devices and the Marvel

People often want to know whether or not brain devices work. This chapter argued that scientific credibility partly depends on how a technique is demonstrated. It analyzed the histories and contemporary uses of several brain devices. Light and sound machines, for example, are well demonstrated in the personal, but not in the public and polemic domain. Promoters of electric and magnetic brain therapies carefully demarcate their practices, but their therapeutic effects are not well enough represented in the polemic domain to achieve scientific credibility. And neurofeedback experts have problems in reenacting their therapeutic demonstrations out of the personal (spiritual, self-help) domain and into a formal experimental setting, and hence to defend them as 'evidence-based' in the polemic domain. In spite of these difficulties, this book shows that this does not mean that these devices do not work. On the contrary, I argue that these devices have effects as long as they are used.

As the introduction quote of this chapter revealed, Lord Kelvin, an important representative of 19th century physics, was able to reenact an effect of 'nothing' as a marvel. A comparable transformation can also occur by using brain therapies. One client who is disappointed about a specific neurofeedback therapy illustrates this: 'So, what did [clinic X] do with my EEG, in 24 sessions and 3500 euro? Well, nothing. And that's what I blame them for. (…) Their answer to why my EEG remained unchanged was 'You have a stubborn brain' (12). In both situations, an experience of nothing was reconstituted into something remarkable, a marvel and a stubborn brain. In the case of Lord Kelvin it is difficult to find out which effect his observation had, but in the example of clinic X the subject sighs after he expressed his anger: 'Well, maybe I do have a stubborn EEG that doesn't want to be changed with their methods, but will do with other methods.' That is to say, although clinic X did 'nothing' for their client's EEG (according to the client), their techniques and explications had certain effects in the sense that this person went on searching for other ways to change his EEG.

Trying to change your EEG in order to change yourself is based on a very different idea of oneself than, for example, going to a psychotherapist to work on your early childhood, or making confessions to purify your soul. As Foucault (1988) demonstrated with his concept of 'technologies of the self', depending on how people think about themselves and their behavior, they will rely on different techniques to work on themselves to improve their behavior, feelings, or selves. This also works the other way around; different techniques to work on oneself are (or can be) based on different knowledge and precepts, and, as a result, people using these techniques will constitute themselves in a different way.

Inspired by Foucault, this book analyzes what kind of subjectivity is constituted by using a brain device. I decided to focus my study on neurofeedback users, since neurofeedback is the most used and promoted of the brain devices that I described. The use of neurofeedback has rapidly increased in Europe and the USA and there is plenty of information available (studies, clinics, clients, practitioners, forums). Another reason to concentrate on neurofeedback is that I expected a larger impact on the self than from light and sound machines or electric or magnetic devices, because doing neurofeedback literally confronts people with their brain activity and directly asks them to intervene in this activity. Hence, I assume that neurofeedback is a technology that changes the self—perhaps not in the sense that it cures or improves the users' brains, like practitioners claim—but in the sense that doing neurofeedback creates new ideas about someone's self, brain, problems, history, and future.

That is, whereas this chapter discussed how practitioners and other representatives do their best to prove the marvelous effects of their devices, below I demonstrate that indeed something marvelous can be observed: the constitution of a new self for the neurofeedback user. However, before

I discuss the acts and explanations of neurofeedback clients and practitioners (Chaps. 4 and 6), I will first explore how brainwaves and psyches were connected to each other, and hence started to interact in the work of some central figures in the history of neurofeedback. My analysis of the ideas and lives of these brainwave scientists will not provide a fully elaborated description of how academics thought and worked upon the brain, but will give some important insights in how working upon the brain interacts with our ideas of self.[23]

References

Abi-Rached, J. M., & Rose, N. (2010). The birth of the neuromolecular gaze. *History of the Human Sciences, 23*(1), 11–36. doi:10.1177/0952695109352407.

Arns, M., de Ridder, S., Strehl, U., Breteler, M., & Coenen, A. (2009). Efficacy of neurofeedback treatment in ADHD: The effects on inattention, impulsivity and hyperactivity: A meta-analysis. *Clinical EEG and Neuroscience, 40*(3), 180–189. doi:10.1177/155005940904000311.

Ashmore, M., Brown, S. D., & Macmillan, K. (2005). Lost in the mall with Mesmer and Wundt: Demarcations and demonstrations in the psychologies. *Science, Technology and Human Values, 30*(1), 76–110. doi:10.1177/0162243904270716.

Brunoni, A. R., Nitsche, M. A., Bolognini, N., Bikson, M., Wagner, T., Merabet, L., et al. (2012). Clinical research with transcranial direct current stimulation (tDCS): Challenges and future directions. *Brain Stimulation, 5*(3), 175–195. doi:10.1016/j.brs.2011.03.002.

Canales, J. (2011). "A number of scenes in a badly cut film": Observation in the age of strobe. In L. Daston & E. Lunbeck (Eds.), *Histories of scientific observation* (pp. 230–254). Chicago/London: The University of Chicago Press.

Coben, R., Linden, M., & Myers, T. E. (2010). Neurofeedback for autistic spectrum disorder: A review of the literature. *Applied Psychophysiology and Biofeedback, 35*(1), 83–105.

de la Peña, C. T. (2001). Designing the electric body: Sexuality, masculinity and the electric belt in America, 1880–1920. *Journal of Design History, 4*, 275–290.

---

[23] That is, this historical exploration (Chap. 3) can be read as a genealogy in the sense of Foucault. A genealogy is 'to discover that truth or being does not lie at the root of what we know and what we are, but the exteriority of accidents' (Foucault, 1984, p. 81). It does not analyze phenomena as inevitable, but tries to understand or diagnose the present by treating their emergence as a question or problem (Abi-Rached & Rose, 2010).

de Rijcke, S., & Beaulieu, A. (2007). Essay review: Taking a good look at why scientific images don't speak for themselves. *Theory and Psychology, 17*(5), 733–742. doi:10.1177/0959354307081626.

Demos, J. N. (2005). *Getting started with neurofeedback.* New York/London: W.W. Norton.

Dumit, J. (2004). *Picturing personhood: Brain scans and biomedical identity* (Information series). Princeton, NJ/Oxford: Princeton University Press.

Ellison, K. (2010). Neurofeedback gains popularity and second looks. *The New York Times.* Retrieved from http://www.nytimes.com/2010/10/05/health/05neurofeedback.html

Foucault, M. (1984). Nietzsche, genealogy, history. In P. Rabinow (Ed.), *The Foucault reader* (pp. 76–100). New York: Pantheon Books.

Foucault, M. (1988). Technologies of the self. In L. M. Martin, H. Gutman, & P. H. Hutton (Eds.), *Technologies of the self. A seminar with Michel Foucault* (pp. 16–49). Amherst: The University of Massachusetts Press.

Geiger, J. (2003). *Chapel of extreme experience. A short history of stroboscopic light and the dream machine.* Brooklyn, NY: Soft Skull Press.

Green, E., & Green, A. (1978). *Beyond biofeedback.* New York: Dell.

Greenwood, J. D. (1996). Freud's "tally" argument, placebo control treatments, and the evaluation of psychotherapy. *Philosophy of Science, 63*(4), 605–621. doi:10.1086/289979.

Gruzelier, J., Egner, T., & Vernon, D. (2006). Validating the efficacy of neurofeedback for optimising performance. *Progress in Brain Research, 159*, 421–431.

Hayward, R. (2001). The tortoise and the love-machine: Grey Walter and the politics of electroencephalography. *Science in Context, 14*(4), 615–641.

Hesse, M. (2010). High fidelity; are kids catching a buzz just by listening to music? I-Doser says fer sure, man. Retrieved from http://www.highbeam.com/doc/1P2-25452146.html

Higgins, E. S., & George, M. S. (2009). *Brain stimulation therapies for clinicians.* Washington, DC: American Psychiatric Publishing.

Horvath, J., Perez, J., Forrow, L., Fregni, F., & Pascual-Leone, A. (2011). Transcranial magnetic stimulation: A historical evaluation and future prognosis of therapeutically relevant ethical concerns. *Journal of Medical Ethics, 37*(3), 137–143. doi:10.1136/jme.2010.039966.

Huitema, R., & Eling, P. A. T. M. (2009). Neurofeedback—Wat is het waard? *Tijdschrift voor Orthopedagogiek, 48*, 115–126.

Hutchison, M. (1990). The megabrain report special issue on sound and light technologies Megabrain Report (Vol. 1). Sausalito, CA: Megabrain Inc, from http://mindplacesupport.com/files/3514/0744/0313/MegaBrain_Report_Volume_1_Number_2.pdf

Huxley, A. (1994). *The doors of perception. Heaven and hell.* London: Flamingo.

Ioannidis, J. P. (2008). Effectiveness of antidepressants: An evidence myth constructed from a thousand randomized trials? *Philosophy, Ethics, and Humanities in Medicine, 3,* 14. doi:10.1186/1747-5341-3-14.

Kamiya, J. (1968). Conscious control of brain waves. *Psychology Today, 1*(11), 56–60.

Loeb, L. (1999). Consumerism and commercial electrotherapy: The medical battery company in nineteenth-century London. *Journal of Victorian Culture, 4*(2), 252–275. doi:10.1080/13555509909505992.

Loftus, E., & Pickrell, J. (1995). The formation of false memories. *Psychiatric Annals, 25*(12), 720–725.

Logemann, H. N. A., Lansbergen, M. M., Van Os, T. W. D. P., Böcker, K. B. E., & Kenemans, J. L. (2010). The effectiveness of EEG-feedback on attention, impulsivity and EEG: A sham feedback controlled study. *Neuroscience Letters, 479*(1), 49–53. doi:10.1016/j.neulet.2010.05.026.

Loo, S. K., & Barkley, R. A. (2005). Clinical utility of EEG in attention deficit hyperactivity disorder. *Applied Neuropsychology, 12*(2), 64–76.

Lubar, J. F., & Shouse, M. N. (1976). EEG and behavioral changes in a hyperkinetic child concurrent with training of the sensorimotor rhythm (SMR). *Biofeedback and Self-Regulation, 1*(3), 293–306.

Macrae, F. (2008). The "thinking cap" that could unlock your inner genius and boost creativity. *Mail Online.* Retrieved from http://www.dailymail.co.uk/sciencetech/article-1064431/The-thinking-cap-unlock-inner-genius-boost-creativity.html

McGoey, L. (2010). Profitable failure: Antidepressant drugs and the triumph of flawed experiments. *History of the Human Sciences, 23*(1), 58–78. doi:10.1177/0952695109352414.

Neuronetics. (2008). *Neurostar TMS therapy* (FDA approval).

Nitsche, M. A., & Paulus, W. (2011). Transcranial direct current stimulation—Update 2011. *Restorative Neurology and Neuroscience, 29*(6), 463–492. doi:10.3233/RNN-2011-0618.

Pascual-Leone, A., & Wagner, T. (2007). A brief summary of the history of noninvasive brain stimulation. *Supplemental Material: Annual Review of Biomedical Engineering.* Retrieved from http://www.annualreviews.org/article/suppl/10.1146/annurev.bioeng.9.061206.133100?file=SupplementalApendix.pdf

Rosch, P. J. (2009). Bioelectromagnetic and subtle energy medicine. *Annals of the New York Academy of Sciences, 1172*(1), 297–311. doi:10.1111/j.1749-6632.2009.04535.x.

Smythies, J. R. (1959a). The stroboscopic patterns. 1. The dark phase. *British Journal of Psychology, 50*(2), 106–116.

Smythies, J. R. (1959b). The stroboscopic patterns.2. The phenomenology of the bright phase and afterimages. *British Journal of Psychology, 50*(4), 305–324.

Smythies, J. R. (1960). The stroboscopic patterns.3. Further experiments and discussion. *British Journal of Psychology, 51*(3), 247–255.

Stengers, I. (1997). *Power and invention: Situating science.* Minneapolis, MN: University of Minnesota Press.

Sterman, M. B., & Macdonald, L. R. (1978). Effects of central cortical EEG feedback training on incidence of poorly controlled seizures. *Epilepsia, 19,* 207–222.

Szasz, T. (2006). The pretense of psychology as science: The myth of mental illness in Statu Nascendi. *Current Psychology, 25*(1), 42–49. doi:10.1007/s12144-006-1015-3.

Tanner, J. M., & Inhelder, B. (1971). *Discussions on child development. In one volume.* The proceedings of the meeting of the World Health Organization Study Group on the Psychobiological Development of the Child, Geneva 1953–1956 (Vols. 1–4, Vol. 1–4). Edinburgh: Tavistock Publications.

Thomson, W. S. (1889). The six gateways of knowledge. *Popular lectures and adresses. 1. constitution of matter,* Nature series (Vols. 1–3, Vol. 1, pp. 253–299). London: MacMillan & Co. Retrieved from http://dss-edit.com/ethernity/Thompson-Kelvin_Popular_Lectures_and_Addresses_Constitut.pdf

van As, J., Hummelen, J. W., & Buitelaar, J. K. (2010). Neurofeedback and attention deficit hyperactivity disorder: What is it and is it working? [Article in Dutch]. *Tijdschrift voor Psychiatrie, 52*(1), 41–50.

Vijselaar, J. (2007). *Psyche en elektriciteit.* Utrecht, The Netherlands: Universiteit Utrecht, Faculteit Geesteswetenschappen.

Vollebregt, M. A., van Dongen-Boomsma, M., Buitelaar, J. K., & Slaats-Willemse, D. (2014). Does EEG-neurofeedback improve neurocognitive functioning in children with attention-deficit/hyperactivity disorder? A systematic review and a double-blind placebo-controlled study. *Journal of Child Psychology and Psychiatry, 55*(5), 460–472.

Walter, W. G. (1953). *The living brain.* London: G. Duckworth.

Walter, W. G. (1956). Comments on professor Piaget's paper. In J. M. Tanner & B. Inheleder (Eds.), *Discussions on child development* (Vol. 4, pp. 53–60). London: Tavistock Publications.

Walter, W. G. (1957). *The living brain.* London: Gerald Duckworth & Co. Ltd.

Wyrwicka, W., & Sterman, M. B. (1968). Instrumental conditioning of sensorimotor cortex EEG spindles in the waking cat. *Physiology and Behavior, 3*(5), 703–707. doi:10.1016/0031-9384(68)90139-X.

3

Glancing Behind the Scenes

In spite of all failures and defeats concerning the secrets that surround our own psyche, of which some will be insoluble forever, the hope to glance behind the scenes has time and time again driven the examining human spirit to new action, because nothing attracts more than the unknown that daily surrounds us. (Berger, 1940, p. 32)

With this message, the German psychiatrist and psychophysiologist Hans Berger, also known as the discoverer of the human electroen-cephalogram (EEG), finished his last published document. The quote represents Berger's lifelong mission and enduring frustration to retrieve the physical equivalent of psychological processes, but Berger makes his mission more general. He predicts that people will never be able to solve the secrets of the human psyche, and he states that these secrets have continuously driven 'the examining human spirit' to new action. In the course of time, many more researchers indeed tried to 'glance behind the scenes', by studying the brain, and according to some optimistic book titles—*Consciousness Explained* (Dennett, 1991), *Explaining the Brain* (Craver, 2009), *Understanding Consciousness* (Velmans, 2000), *Brain-wise* (Churchland, 2002), *Self Comes to Mind* (Damasio, 2012),

J. Brenninkmeijer, *Neurotechnologies of the Self*,
DOI 10.1057/978-1-137-53386-9_3

Synaptic Self (LeDoux, 2003)—with more success (or exaggeration) than Berger.

These book titles suggest to have solved 'the secrets that surround our own psyche' by studying the self or consciousness as part of the brain. They all proclaim a reductionist vision of the self which evoked the idea that improving oneself could be achieved by changing one's brain. However, as argued in the introduction, working on the self by working on the brain gives rise to an ontological difficulty since the self becomes the object and the subject of the brain, at the same time. That is, the self is the operator who studies or trains the brain, but also the product of or even the same entity as this brain. To understand how brain device users deal with this ambiguity, I thoroughly describe their practices and explanations while doing neurofeedback in Chaps. 4 and 6. By choosing neurofeedback to improve oneself, people suggest that these selves are, or are in, their brains, but while doing neurofeedback they have to distinguish themselves from their brains in order to make it possible to act on their brains. In other words, working on the self by working on the brain might be inspired by a reductionist view, but the act itself basically assumes an interaction of two actors.

To understand how this interaction between brains and selves emerged, this chapter explores the work and philosophies of early brainwave scientists and contemporary neurofeedback experts. I studied the academic work, diary notes, and media appearances of four scientists who are often mentioned in the histories of clinical EEG and neurofeedback (e.g. Robbins, 2000; Budzynski, 1999; Demos, 2005; Niedermeyer & Lopes da Silva, 2005), and compared these reports with the acts and explanations of contemporary neurofeedback experts (Fig. 3.1). I analyzed how human brainwaves have been associated with something like the self, or the psyche, since their first demonstration by Hans Berger (1873–1941), how they were subsequently connected to personality types by the British neurophysiologist and cyberneticist William Grey Walter (1910–1977), and how they were made trainable by the American psychologists Joe Kamiya (1925–) and Barry Sterman (1935–). Furthermore, I show how contemporary neurofeedback practitioners learn to focus on the brain instead of the mind but also struggle with the transformation from psychology to physiology. My analysis will demonstrate that working on the self by working on the brain is often incited by personal experiences or spiritual beliefs, and constitutes an increasingly complicated relationship between the brain and the self.

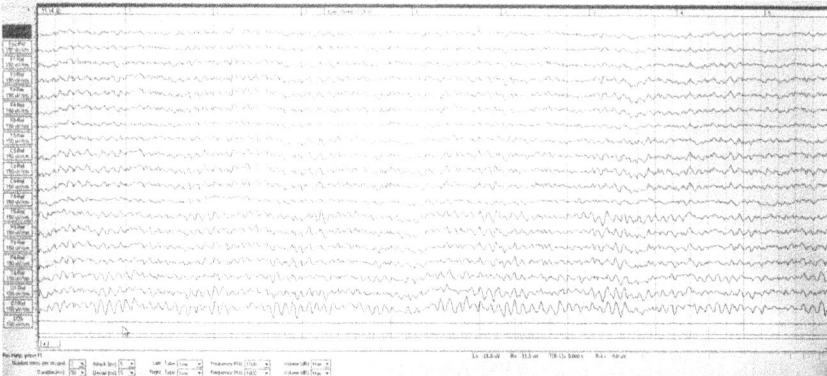

Fig. 3.1 Human brainwaves (a raw EEG) recorded in a neurofeedback clinic (Used with permission of Roland Verment)

The Ungraspable Psyche

The demonstration of the human electroencephalogram was not a matter of course. It took Hans Berger thirty years of recording brain activity before he dared to publish the sentence which would make him famous: 'I therefore, indeed, believe that I have discovered the electroencephalogram of man and that I have published it here for the first time' (Berger 1929, as published in Berger, 1969, p. 70). During his entire career Berger struggled with a personal mission: he wanted to prove the existence of psychical energy. This mission was not a result of his neurophysiological findings, but preceded his scientific career. According to one of his most prominent biographers, Berger was already 'absorbed by the mind–body problem' in his teenage years (Gloor, 1994, p. 253).

One important episode for his devotion was 'a case of spontaneous telepathy' that Berger retrospectively described in his last published document, named *Psyche*:

As a 19 year old student, I had a serious accident during a military exercise near Würzburg and barely escaped certain death. (…) In the evening of the same day, I received a telegram from my father who enquired about my well being. (…) This is a case of spontaneous telepathy in which at a time of mortal danger, and as I contemplated certain death, I transmitted my thoughts, while my sister, who was particularly close to me, acted as the receiver. (Berger, 1940, pp. 5, 6; translated by Gloor, 1969, pp. 2, 3)

Likely, this telepathic experience initiated Berger's decision to explore the relation between physical and psychical events, since shortly after the accident he changed his studies from astronomy to medicine (Gloor, 1969).

Berger spent his entire career at the Jena Psychiatric University Clinic, from his doctoral degree in 1897 until his retirement in 1938. The first few years he did some work in neuroanatomy, but his drive to find a connection between mental and physical events soon emerged in his research. In 1901 he published his first psychophysiological experiments about the blood volume changes in the brain of a trepanned patient who was given pharmaceuticals, like chloroform, cocaine, morphine, and amyl nitrite, that influence mental activity. Furthermore, he studied changes of the bloodstream in the brain during various 'psychological' states such as attention, affects, and sensory stimuli (Gloor, 1969, p. 4), measured people's brain temperature during different mental conditions (Berger, 1910), and attempted to detect mental conditions in the blood, for instance, by injecting himself with blood samples of psychotic patients to see if the disease was transferable by blood (Boening, 1941; Borck, 2005a). None of these experiments, however, produced any revolutionary results and they evoked little scientific response (Jung, 1963).

In the same period, Berger made the first attempts to record electric brain activity in animals and later in humans with trepanated or broken skulls. However, these experiments were also disappointing, since they did not reveal any clear results. In 1910, thirteen years after he started his career, and eight years after the first attempt to record a brain signal, Berger expressed his frustration about the ungraspable human brain recordings in his diaries: 'Of nine experiments, one success and even this one rather doubtful (…). One can therefore not say that I gave this thing up lightly. Eight years! Trying always, time and again' (Gloor, 1969, p. 5).

Despite these disappointments, Berger did not set aside his mission, but accentuated it: 'Psychical Energy is the major challenge! Especially assigned to me' (Borck, 2005a, p. 76). Apparently, Berger was so eager to solve this personal assignment that he did not involve any colleagues in his work but practiced his experiments solely after working hours, as if he had a secret mission (Ginzberg, 1949, p. 364). However, the many failed experiments, perhaps combined with the fact that Berger worked completely on his own, made him somewhat uncertain about the few results he did find. His first successful EEG recording from a non-trepanated skull was made in 1924 of his son Klaus, but Berger made many more EEGs before he was convinced that the measured activity was really brain activity rather than artifacts from the machine or muscle movements (Spear, 2004). It took Berger five more years before he dared to publish his results about the 'writings of the human brain', which he called electroencephalograms, in analogy to the human electrocardiogram (Berger, 1969; Borck, 2001). Moreover, even for this report, which would be followed up by 13 more, Berger had decided 'not to go into hypothetical matters with the publication on the EEG, but only communicate purely concrete facts and findings!' (Millett, 2001, pp. 537, 538; Berger's diary 1929).

The next step was to reveal the meaning of these recordings. One of the first things Berger noticed was a difference in the EEG when people opened or closed their eyes, as well as between subjects doing nothing or performing mental tasks:

> I had been struck early by the fact that in many experimental subjects opening of the eyes, while recording the curve from the skull surface, caused an immediate change in the EEG and that during mental tasks, e.g., when solving a problem of arithmetic, the mere naming of the task sometimes caused the same change of the EEG. (Berger, 1969, p. 83)

These changes appeared in specific wave patterns that intensified or reduced. The first pattern Berger determined he called the alpha rhythm, also described by him as 'the physical concomitants of conscious phenomena' (Millett, 2001, p. 539) and as corresponding with 'fluctuations of attention' (Berger, 1969, p. 79). Berger further connected these brain fluctuations with the subject's mental condition by demonstrating that anxiety or attention can change the EEG, and by comparing EEGs with people's intelligence. In his 14th report Berger, for example, reported: 'One may even observe the peculiar fact that mental defectives in general exhibit better resting EEG curves than intelligent persons. When I wanted to demonstrate beautiful EEGs to colleagues who were interested in such recordings, I particularly liked to use a certain imbecile' (Berger, 1969, pp. 315, 316).

Berger connected brainwaves to mental activity, but he was rather careful in interpreting his EEG results and going into 'hypothetical matters'. In spite of this caution, or perhaps because of it, Berger's results were nearly completely ignored in the academic world. Only five years after publication, Berger's brain rhythms were confirmed by the British neurophysiologists Edgar Adrian and Bryan Matthews (Adrian & Matthews, 1934). Although these scientists were rather skeptical about Berger's interpretations—they, for example, demonstrated that the EEG of Adrian had more similarities with that of a water beetle than with the EEG of his colleague Matthews—it gave Berger some of the recognition he was waiting for. After this publication Berger was invited to be a co-chairman with Adrian at a symposium on electrical activity in the nervous system and was hailed as the most 'distinguished of all visitors', at which occasion a colleague made the observation that 'tears came to his eyes as he replied, "In Germany I am not so famous"' (Gibbs, 1941, p. 516).

One of the reasons why Berger was not so famous, at least in Germany, might have been his topic of research. According to Pierre Gloor, Berger's main translator and biographer, Berger chose the 'difficult role of being an outsider' with his psychophysiological interest (Gloor, 1969, p. 3). Although the field of psychophysiology had been flourishing in the 19th century with the work of Weber, Fechner, Helmholtz, Hering, and Wundt, it had fallen into 'disrepute among neurologists and psychiatrists' (Gloor, 1969, p. 3) at the end of the century. A neuroanatomical approach (with work of Gudden, Meynert, Flechsig, Forel, and von Monakow) and a functional approach (with work of Kraepelin, Bleuler, Janet, Freud, and Adler) had become fashionable, but Berger did not join either of these. Instead, he drew his inspiration from the electrophysiological experiments on animals from Caton, Fleisch von Marxow, Beck, and Cybulski (Berger, 1969, pp. 37–38; Gloor, 1969, p. 4). Most scientists in that period, however, simply did not believe in electrical measurements of the brain and considered the (weak) electrical oscillations as artifacts of the apparatuses (Jung, 1963; Spear, 2004).

If it was for this lack of recognition or for other reasons, the story goes that one day in 1938 Berger was suddenly informed that his retirement would start the next morning.[1] Berger was no longer welcome at the institute where he had worked for almost forty years, quietly experimenting and gathering his results. The precise effects of the end of his career, combined with the frustration about the lack of scientific recognition and perhaps the start of the war, are difficult to reconstruct, but according to his biographers,[2] Berger's last years were rather tragic. In his final work *Psyche* (1940), he revealed his ideas about psychical energy and telepathy, but he combined these philosophical and parapsychological interests with a rather physical understanding of his own psyche. According to his biographers, Berger became increasingly despondent, but did not recognize (or accept) the psychiatric character of his problems. Under the notion of having heart problems, Berger went to the medical clinic in Jena in May 1941, where he hung himself (Brazier, 1961; Gloor, 1969; Jung, 1963).[3]

Brain and Soul

Berger is nowadays known as the 'discoverer of human EEG', but given the time, frustration, and efforts he invested, this 'discovery' is better considered an invention achieved by hard work. Berger created the human EEG because he had a personal quest. His ideas about 'Psychische Energie' did not develop spontaneously after his 'discovery' of the human EEG; it was the other way around. He pursued an electric pattern in the brain because he was searching for evidence of the existence of psychical energy. That is to say, Berger's neuroscientific investigations were incited by a personal and somewhat spiritual quest.

---

[1] Borck (2005a), however, refutes this rumor and argues that Berger was aware of his retirement before, which was at the regular age of 65.

[2] Berger actually lacks a biographer. Gloor wrote a short introduction about Berger's life in his translation of Berger's work (Gloor, 1969); Jung published most of his diary fragments (Jung, 1963; Jung & Berger, 1979); and Berger's colleague Ginzberg wrote a 'contribution to his biography' (Ginzberg, 1949).

[3] To make it even more dramatic; in the same year that Berger died, his son Klaus—the subject whose brain was first visualized with Berger's EEG—also passed away.

This spiritual connotation is omnipresent in his final work *Psyche* (Berger 1940) in which Berger explains his ideas about telepathy and psychical energy, but also in his diary where Berger regularly refers to his investigations as a 'personal assignment', and sometimes describes his finding of the human EEG by thanking God: 'I have just recorded splendid electroencephalograms with chlorinated silver needles! I thank you, my God!' (Borck, 2005a, p. 82, Berger diary 1929). Besides this spiritual touch, however, Berger holds a materialistic position. He occasionally refers to the monistic and pantheistic philosopher Spinoza who stated that 'God' is a self-subsistent substance of which mind and matter are qualities (de Spinoza, 2006). Berger states in one of his diaries: 'My god is the god of Plato, Spinoza, Goethe and others' (Jung, 1963, p. 28). The convergence of these references leads to the conclusion that Berger's mission can be seen as a spiritual mission in the sense that his attempts to reveal the connection between psychical and physical phenomena can be interpreted as an attempt to comprehend God (c.f. Millett, 2001).

Berger is nowadays often mentioned as an important neuroscientist and a monistic thinker (Gloor, 1994; Millett, 2001). In several texts, Berger emphasizes the homogeneity of psychophysiological activity. In his 6th, 11th, and 14th report, for example, he writes that psychophysiological activity in the cortex acts as 'ein einheitliches Ganzes', which can be translated as a 'uniform' or 'homogeneous' whole. In his inaugural lecture 'Brain and soul' (1919),[4] delivered on the occasion of his appointment as director of the psychiatric clinic in succession to Otto Binswanger, Berger explains:

> I openly declare that I do not hold the popular parallel principle as the solution [to the mind–brain problem], but instead I accept an interaction between mental and bodily processes and embrace an energetic perspective, against which all possible objections can be raised, like any other assumption. (Millett, 2001, p. 533)

In this quotation, Berger seems to oppose dualism, but one can wonder if an interaction between mind and body, although described from

---

[4] Berger (1919) Hirn und Seele, Fischer: Jena. Unfortunately, I had no access to the document.

an energetic perspective, should be considered as a monistic statement. In his final work *Psyche*, Berger expresses this duality of an interacting oneness:

> This psychical energy (…) fundamentally distinguishes itself from all other kinds of energies, but can interact with, or rather arise from, and retransform into these. One can rightly argue against this assumption that it maintains the old Dualism of material and psychical processes, only in a somewhat concealed form. This can be admitted easily and does the view no harm. (Berger, 1940, p. 24)

Berger used 'the law of the conservation of energy'[5] to connect physical energy with psychical phenomena, but, as he states himself, this does not make him a confirmed monistic thinker. Moreover, the German historian Cornelius Borck even argues: 'He was a dualist, and he sought to fight materialism with its own weapons' (Borck, 2005b, p. 83). That is, apparently the line between monism and dualism is not always as clear as it is often claimed.

It is indefinite whether Berger should be considered as a monistic, a dualistic, or maybe even a holistic thinker (see also Borck, 2001),[6] but what can be concluded is that the connection between mind and body started as Berger's mission turned into his frustration, and resulted in a feeling of failure. In his final work *Psyche* Berger declares that 'it is absolutely sure that it will never be completely revealed how material processes of the cerebral cortex and the corresponding psychical processes will be related in the end' (Berger, 1940, p. 16). One could even argue that the ungraspable psyche which was such a drive and struggle during Berger's life finally brought him down, since by committing suicide Berger was actually overrun by his own psyche.

---

[5] Several other scientists connected the 'law of conservation of energy' to psychological phenomena. Berger was, for example, inspired by Alfred Lehman (1858–1921). See Sourkes (2006) for more information about the consequences of the law of conservation of energy for psychological theories.

[6] Holism understands systems or organisms as a whole, instead of a sum of elements. Physiological processes are seen in terms of their roles in the total functioning of the organism, and mind and body are not ontologically different. For holistic thinkers, the mind is in the body, and the body is reanimated with a mind. (Harrington, 1999)

The scientific impact of Berger's work took some time and predominantly occurred after his death, but some impact in society started immediately following Berger's first EEG publication in 1929. Berger's 'brain mirror' gained attention in German newspapers, which described the invention as producing the 'zig-zag line of the human soul' and the 'electric script of thinking' (Borck, 2001). According to the German press, the EEG was not only a recording, but also a deciphering of the language of the operating brain, and they made Berger the inventor who mastered the brain 'to write in black on white' (Borck, 2001, p. 584), by publishing his picture besides samples of his recordings. Whilst it is difficult to track the precise impact of these articles that reached a broad public, it is presumable that the first confrontation with a brain script that is translated as the visualization of the human soul or the inner voice had some impact on people's understanding of their selves.

Brain Brothers

The British neurophysiologist and cyberneticist William Grey Walter was one of Berger's followers. However, he was not particularly flattering about his predecessor. In his famous book *The Living Brain* (Walter, 1953a, 1957) Walter described Berger as 'a surprisingly unscientific scientist', with the 'reputation of a crank' being 'completely ignorant of the technical and physical basis of his method'. Furthermore, he referred to Berger's recordings as 'wobbly line [s]' that 'did not convince us or anybody else at that time' (Walter, 1957, pp. 29–30). However, at the insistence of his laboratory director at the Maudsley Mental Hospital in London, the psychiatrist Frederic Golla, Walter did get involved in EEG practices in 1934 (Hayward, 2001; Pickering, 2010). Golla sought to find 'a way from the dead world of science to the living world of purposes and values' (Golla 1938, as cited in Hayward, 2001, p. 619) and he needed the physiological experience of Walter—who had worked as an undergraduate for Adrian and Matthews in Cambridge—to understand the working of the EEG.

First at the Maudsley mental hospital and since 1939 at the BNI in Bristol, Walter and Golla performed many EEG experiments together.

They confirmed Berger's alpha waves and soon traced a new rhythm, which they named delta, because of its association with 'disease, degeneration and death' (Walter, 1957, p. 53). The delta wave appeared to be usable for the detection of cerebral brain tumors and epilepsy, and, shortly after its detection, Walter used the EEG in the defense of a man who murdered a schoolgirl, by 'showing' that the defendant was an epileptic who had attacked the girl during a seizure. This story attracted much attention in the British press and gave the EEG the status of some kind of 'truth machine or electric confessional that would reveal the occult working of the human mind' (Hayward, 2001, p. 620).

Walter's detection of the delta wave made the EEG an important diagnostic tool in medical sciences, and his later detections of the theta wave and the contingent negative variation (CNV)[7] gave him an important role in science (Hayward, 2001; Holland, 2003). The alpha wave, on the other hand, led to interesting speculations about brains and personality types. When performing his EEG experiments, Walter had soon found out that not every brain produced the same patterns of alpha. His own brain, for example, did not produce any alpha, and other brains produced the rhythm all the time, while most brains produced alpha only with closed eyes and 'a blank mind'. Walter concluded that different alpha activities should correspond to different ways of thinking, and thus to different personalities. He developed a theory of three different alpha types. The M type stood for minus, no alpha, and this person was an image-based thinker like himself. The P type (persistence alpha) was an abstract thinker, and the R type (regular alpha) could switch between abstract and image-based thinking.

To identify someone's alpha type one could measure the subject's brainwaves, or give a mental exercise in which a painted cube had to be halved several times. After getting the puzzle, the subject was questioned about the colors and structure of the cube. During a lecture Walter declaimed:

> How many of you, I wonder, saw not merely a color but the grain of the wood, perhaps the knife or saw-blade or sawdust? These I would hail as my

---

[7] The CNV, also called 'the expectancy wave', and one of the first event-related potentials (ERP) described, is the EEG component that is measured between the expectation of and reaction on a stimulus.

brain-brothers. I would expect that to some extent they have followed my arguments, shared my images, even if they have not agreed with me. Those who saw nothing (...), the ones who computed without form or color, I salute as distinguished strangers in my brain country, I fear they may have found my examples trivial and my arguments tedious even if they do agree with me. (Walter, 1969a, p. 23)

Walter categorized people under M, P, and R types, and talked about brain brothers who would understand each other and brain strangers who might think and communicate differently. In other texts, he claimed that differences in alpha waves could not only cause miscommunications between people, but also serious problems in marriages, science, society, and the world in general:

Their mental accents, so to say, separate them as surely as verbal accents in a class-conscious society. Of course, it is not only among scientists that such discrepancies can cause irrational rupture of communication. (...) It may even be that serious crises between nations (...) have arisen because the negotiators have different types of imagery and can only talk at cross-purposes. Some current conflicts that threaten to tear our world asunder may be no more serious in origin than an argument about whether the cube was 'really' red or blue. Perhaps a diplomat should have his alpha-type endorsed on his passport. (Walter, 1968a, p. 184)

By categorizing people under different alpha types Walter presumed that people have a 'fixed' brain state, or brain personality. He referred to EEGs as 'brainprints', in analogy to fingerprints, and he claimed that although the EEG changes continually its trends are very individual and identifiable (Walter, 1957, p. 136). On at least two occasions, however, his own life experiences contradicted his alpha theories. After he divorced his wife—who had a similar alpha type—and fell in love with a woman with a different alpha type, Walter changed his view about couples who could only match if their brains would match by adding that different alpha types could also complement each other's ideas (Hayward, 2001). And when one of his colleagues measured alpha waves in Walter's brain after a brain injury, caused by a motor accident, Walter adapted his personality—

which theoretically had been changed from an M type into an R type—to his temporarily changed brain: 'I recalled that around the period of my alpha rhythm my mind seemed capable of "free-wheeling"—feeling blank but healthy, which was a novel experience to me. Later, my visual images began to return obtrusively and now I feel quite like I remember before the injury' (Walter, 1972a, p. 48).

His brain type theories are furthermore contradicted with his explanations of the theta waves. Walter associated theta, a rhythm dominant in infants, with pleasure and pain and, when prominent in (bad-tempered) adults, with 'childish' behavior like intolerance, selfishness, impatience, and suspicion (Walter, 1957, pp. 140, 144). Being able to control these theta waves, on the other hand, he associated with self-control, personality, and maturity. In several texts (Walter, 1952, 1957, 1960) Walter described experiments in which subjects were stimulated with flickering lights of theta frequency. This stimulus provoked theta waves in the brain, as Walter illustrated with an EEG, and was supposed to arouse an annoying feeling. However, according to Walter people were not 'at the mercy of' their theta rhythms (Walter, 1957, p. 146). Subjects who produced theta and who were confronted with a bad or annoying feeling would consequently (or if they were told to) try to suppress this feeling, and, by successfully repressing this emotion, their theta waves would decrease. In Walter's words: 'If he [the subject] gives way to his feelings, the theta pattern will increase, but if he tries to keep his temper, the pattern fades away, and so does the feeling of annoyance' (Walter, 1952). That is, in this explanation, Walter describes an interaction between a brain and a subject that comes close to the act of doing neurofeedback.

Walter gave rise to a controlling brain with his alpha theories, but with his theta experiments, he conjured up the image of a controllable brain. In other words, the brain is both static and plastic, determining and manipulable, and dominant and obedient. And in both situations (controlling alpha and controllable theta) the brain appears to interact with the subject—it actively determines someone's personality type, but passively decreases its theta waves when the subject tries to keep one's temper.

Technopolis

Walter realized that his work contained some dichotomies, but according to him such distinctions between the brain and the self could only be linguistic. He explained: 'I suppose it is always possible to define one's observations in such a way as to permit a dichotomy, and this may be operationally useful as long as one remembers that it is a descriptive device not an explanation' (Walter, 1971, p. 45). In a number of texts he claimed physiological unity (Walter, 1953b) and he called himself a 'thorough-going materialist' (Walter, 1971, p. 48). Walter furthermore claimed he was completely uninterested in notions of autonomy or identity (Walter, 1957) and he explained the relation between mind and brain as velocity versus the engine (Hayward, 2001; Walter, 1957). Instead of working with the concept of mind, he preferred using the word mentality, as being a brain function.

Surprisingly enough, however, Walter had no problems combining these strictly materialistic ideas with spiritual phenomena. *The Living Brain* (1953a) on the one hand is a 'down-to-earth, materialist and evolutionary story of how the brain functions' and on the other hand a book full of references to 'dreams, visions, ESP [extrasensory perception], nirvana and the magical powers of the Eastern yogi' (Pickering, 2008, p. 1). For Walter these kind of phenomena were 'physiological curiosities' (1957, p. 175) that were perhaps hard to explain in biological terms, but, nevertheless, attracted his attention. Moreover, on some occasions, there even emerged a sort of homunculus in Walter's theories. He wrote, for example, about 'the notion of an intelligible mechanism even in our own brains' (Walter, 1969b, p. 107), and made notion of 'that "restless beast within our heads" that makes each of us distinct and unique' (Walter, 1968a, p. 179). He also used the metaphor of a traffic control system to grasp this internal reflexivity, or controllability, by adding: 'An important feature of this system is that while the control points check or promote circulation, they are also to a limited extent controlled by the traffic itself' (Walter, 1969a, p. 19).

The answer to studying uncontrollable systems like the brain, and perhaps in the future also the solution to solving the (linguistic) mind–body distinctions, was cybernetics (Walter, 1956, p. 53, 1971). Adherents of this approach studied the communication and control between 'systems', with which they could refer to the human or animal brain, as well as to a computer or machine (Pickering, 2010). In Walter's vision the brain was an organ that adapted to its environment, just like other systems did. To demonstrate the equivalence between systems like brains and machines, he developed robotic tortoises that performed 'human' behavior. These tortoises had photoelectric cells that responded to light and electrical contacts that made them reactive to touch. The photoelectric cells made the tortoise move to light, but because these cells were placed on a rotating motor the tortoises could only move in arcs. Whenever the tortoises touched an object their electrical contacts made them move back again (Hayward, 2001, p. 623). The result of these constructions was a tortoise that could dance on its own in front of a mirror, or have a (somewhat flirty) dance with another tortoise. According to Walter, his tortoises had 'free-will', 'recognised their selves' in a mirror, and organized their behavior in a social way (Walter, 1960).

Studying the human as a self-adjusting machine, and the brain as an adaptive system, was very productive during Walter's career, but the closer he reached the end of it, the more he seemed to worry about the consequences of his ideas. This worry was two-sided. On the one hand, Walter worried about the power of knowledge and its effect on humanity or subjectivity, for example, by stating: 'The danger in EEG is that the proper study of mankind by man will in turn be stripped of human qualities' (Walter, 1968b, pp. 763–764). On the other hand, he worried about the powers of machines that were created by man, but not fully controlled. In one of his texts he warned that electronic computers 'threaten to become the master rather than the slaves' (Walter, 1968c, p. 140), and in another he wrote: 'There is a real danger that the widespread application of computers could result in our being trapped, each of us, in an indissoluble wedlock with a particular system, isolated, house-proud and complacent in the suburbia of technopolis' (Walter, 1971, p. 43).

On some occasions, Walter's ideas and worries about the connection between human and machine became literally embodied in his personal life. He portrayed himself several times in his living room with his robotic tortoises and child crawling around his knees, as being his three children. After he and his wife divorced, Walter gave one of his robots a female erotic character (Hayward, 2001), and when recovering from his motor accident he wrote an article about his hypnagogic fantasies in which his body parts were replaced or changed by surgeons, as if he was a machine with exchangeable parts (Hayward, 2001; Walter, 1972a). When he gradually regained the control of his own body and thinking, he was confronted with a body that was partly stripped of its human qualities: 'I must learn to stand and walk and talk and write and calculate and write programs for our computers and design experiments and ... and ... THINK. How did I think?' (Walter, 1972a, p. 44). He examined himself as a man with emotions that seemed strange to him: 'Sometimes I felt a tear trickle down my cheek. I thought at first it was a sensation due to my brain injury, but it was wet and a little salty so that I could acknowledge my sentiment as deep and pure' (1972a, p. 44).

Walter's scientific theories had a deep impact on his own personal life and ideas, but also influenced the self-conception of his public. With his book *The Living Brain* (1953a), as well as with performances and illustrations in newspapers and magazines, he demonstrated to a broad public that the brain was an entity that could act upon the person. That is to say, he created a 'performative brain' (Pickering, 2010). This brain could define someone's personality type (M, P, R); it could influence success in marriage, struggles with colleagues, and even world peace ('Electronic Patterns of the Brain', 1956; Walter, 1957, 1968a); and it could take over someone's responsibility, as in the case of the epileptic murderer (Hayward, 2001). Moreover, Walter made a clear connection between human and machine, and employed a machine-like version of the brain. Some of his arguments even read as quotations from contemporary brain bestsellers or interview phrases of neurofeedback users. For example Walter's claim that 'the brain has a capacity for resetting itself, for setting up its own wiring' (Walter, 1953b, p. 141), resembles expressions of authors like Doidge (2007) or Amen (1998) who write about rewiring and resetting brains, and could also be compared with phrases of neurofeedback users who talk about a 'defragmentation of your computer', 'a computer wiring me', or 'my system is unstable' (see Chap. 4).

In addition, as already described in Chap. 2, Walter taught people that they—themselves—could act upon their brains. With *The Living Brain*, he informed a broad audience about the possibility of provoking visions, hallucinations, or 'waking dreams' by gazing into a stroboscope with alpha frequency. This inspired several artists and researchers to build their own brain-manipulating flicker machines (Geiger, 2003). That is to say, Walter introduced technologies to work, or experiment, on the brain, and with this on the self. The promises of these technologies were not as extensive as the promises of the devices people use nowadays, but Walter definitely paved the way for some first steps into brain work.

With his detection of theta and delta brainwaves, the connection of these waves to psychical states and personalities, and the demonstration of their controllability, Walter should be considered as an important character in the history of neurofeedback. Moreover, some of his fantasies and worries about the future of his machines and theories are still present. For example, Walter speculated about brainwaves and the future possibility to discover delinquency beforehand: an idea that has never been realized, but is still popular today. Worries about machines that will take over human control or threaten human qualities and cyborg fantasies are also still present, for example, in media, novels, and movies. Moreover, the way he wrote about brain types and the human as a machine also corresponds to statements people make nowadays. That is, Walter actually had a point when he predicted the influence of cybernetics: 'Our art and science can and should influence the way people think about themselves and one another, about what they mean by happiness, and thence how they plan their ways of living' (Walter, 1971, p. 40).

Desirable Alpha

Berger and Walter are important figures in the history of the EEG (e.g. Niedermeyer & Lopes da Silva, 2005), but contemporary neurofeedback practitioners usually identify two other pioneers of their discipline (e.g. Robbins, 2000; Budzynski, 1999; Demos, 2005). One of them is the psychologist Joe Kamiya who taught his human subjects to recognize and manipulate their own brain states. Another frequently mentioned experiment is that of 'the cats' that were trained to produce a specific brain rhythm by the neuropsychologist Barry Sterman. Sterman's and Kamiya's claims were significant. They argued that brainwaves interacted with the subject's will power, personality, and consciousness.

Kamiya was a medical psychologist of the Langley Porter Neuro-psychiatry Institute in San Francisco, and in contrast to most of his con-temporaries—including Walter—Kamiya claimed to be really interested in subjective experiences. In a recent article, Kamiya reflects on the 1950s and 1960s, when he performed several EEG biofeedback experiments with human subjects and motivates: 'For me, such elements of private experience as feelings, images, thoughts, and hopes were a fundamental feature of human life (...). The apparent denial of their relevance for understanding behavior for the sake of scientific rigor seemed self-defeat-ing' (Kamiya, 2011, p. 65). This intention to pay attention to private experiences, hopes, and feelings, however, was put into practice in a way that was shaped by the behaviorist tradition of the 1950s.[8] In his experi-ments, Kamiya attached subjects to an EEG device and frequently asked them to identify, at the sign of a ringing bell, whether they thought they were in brain state 'A' (alpha) or 'B' (no alpha), whereupon he told them whether they were correct. In this way, Kamiya claimed, his subjects were trained to recognize their own brain state (A or B). Furthermore, he trained his subjects to try to suppress and enhance this alpha state, with apparent success (Kamiya, 1968, 1969, 1971).

According to Kamiya, his test subjects 'had learned to read [their] own brain, or [their] mind' (Kamiya, 1971, p. 282). He quoted his sub-jects while being in alpha state with positive phrases like 'pleasantness', 'some kind of relaxation', 'a general calming-down of the mind' (Kamiya, 1971, p. 287, see also Kamiya 1968). According to Kamiya, psychothera-pists and other people who 'are good at intuitively sensing the way you feel' are good alpha controllers, and Kamiya furthermore reported that he 'generally tend[s] to have more positively liking for the individual who subsequently turns out to learn alpha control more readily' (1971, pp. 287, 288). Kamiya described the alpha state as a 'desirable thing' (1971, p. 288) and he connected this desirable state to something spiri-tual by doing experiments on practiced Zen meditators, and by reporting that subjects who were good at controlling their alpha were mostly also interested in meditative or introspective practices (1968).

As he himself explained, Kamiya did not publish much of his work in scientific journals, but he gave many presentations at scientific and 'civic'

---

[8] That is, Kamiya used operant conditioning procedures, but explains that he was also inspired by cognitive psychologists (Kamiya, 2011).

groups (Kamiya, 2011). Furthermore, he published his findings in the popular magazine *Psychology Today* (Kamiya, 1968). In this way, Kamiya showed to a broad public how they could influence their own brain-waves and what pleasurable or spiritual effects this might have. Hence, his alpha trainings became very popular, and resulted in the situation that 'I [Kamiya] no longer pay Ss [subjects], and I have a list a mile long from various people who call me on the telephone or write me from New York, and other places all over the country to ask if they can come over and serve as subjects!' (Kamiya, 1971, p. 288).

Kamiya's experiments became famous, especially in spiritual circles, and his name can still be found in several yoga and meditation books. The connection between brain activity and spirituality was not new in this area of study: Berger was inspired by his telepathic event, Walter was not surprised by all kinds of spiritual effects of the brain, and even Kamiya's contemporary Sterman would admit that 'there is usually a general response to biofeedback training which resembles some aspects of meditational and Yoga experiences' (Sterman, 1981, p. 405). However, Kamiya's simple experiments inspired a general public to try to reach the 'desirable' alpha state by themselves. In a newspaper article that is repeatedly republished, among others in a textbook for students (Zimbardo & Maslach, 1977), Kamiya is portrayed as 'a pop hero to kids who hoped to groove their way into an instant satori' (Luce & Peper, 1971).

Kamiya was the first researcher in the history of neurofeedback who seriously tried to pay attention to the feelings and reports of his subjects. He called it 'self-defeating' to deny the experiences of subjects and he gave some control back to the self, by informing a broad audience about the possibilities to recognize, change, or enhance their brain state. The self that he restored, however, was not only a mechanical self that could be trained by a ringing bell, but also a spiritual self that could try to change one's brain state to a more meditative one.

Brain Control

According to Kamiya, his alpha experiments not only encouraged people to serve as subjects, but also inspired several important scientists (Kamiya, 2011). Apart from the famous behavioral psychologist B.F. Skinner,

Kamiya reports a visit of Sterman, now a professor emeritus of the University of California, Los Angeles (UCLA), and seen as one of the 'founding fathers' of neurofeedback. Sterman completed his PhD in neurology and psychology and started to work as a sleep researcher in the 1960s. Inspired by Pavlov's conditioning experiments with dogs, Sterman put cats in boxes and conditioned them to press a lever for food, after a bell rang. The cats were connected to EEG devices and in this way Sterman observed that, when the cats sat still while waiting for the right moment (3 seconds after the bell) to press a lever for food, they produced a brain rhythm of 12–16 Hertz on the sensorimotor cortex. Probably encouraged by Kamiya's experiments, Sterman thereupon started to train his cats to produce this so-called sensorimotor rhythm (SMR) 'at will' by rewarding the cats with food whenever they showed the brain pattern. After Sterman stopped rewarding his cats they continued producing SMR more than average. Furthermore, Sterman noticed that the sleep EEGs of the cats were altered, and that they slept more soundly and woke up less (Robbins, 2000; Sterman & Egner, 2006; Sterman & Wyrwicka, 1967).

Around the same time, Sterman was asked by the National Aeronautics and Space Administration (NASA) to examine the toxic effects of rocket fuel. He injected 50 cats with the toxicant which promptly evoked epileptic seizures in most of them. Seven cats, however, resisted the seizures somewhat longer, and three of them were not harmed at all. Sterman did not understand what the difference was between the cats, and it took him several years to realize that the seizure-resistant cats were the ones he had trained to produce the SMR rhythms (Robbins, 2000).[9]

Shortly after this finding, Sterman tried his SMR remedy on a secretary: Mary, 'a 23-year-old white female with a history of convulsive disorder'. Sterman trained her for four months and her seizures reduced from an average of two per month (varied from 0 to 3) to only one seizure that appeared after three seizure-free months—this apart from a double

---

[9] This information is based on interviews with Sterman as phrased in Robbins (2000). The original cat study is never published, because it is property of the US Air Force. However, in 2010, Sterman published a reproduced paper of his original cat results and described how the toxic effects of rocket fuel were studied in eight SMR-trained cats (Sterman, LoPresti, & Fairchild, 2010). That is, in contrast to the information as can be read in Robbins (2000), this reproduced article suggests that the original study aimed to find the effects of SMR training.

seizure six days after the first session (Sterman & Friar, 1972). Apart from these results, Sterman noticed that the subject showed changes in her sleep and personality:

> Having previously been a quiet and unobtrusive individual, she progressively became more outgoing, showing increased personal confidence and an enhanced interest in her appearance. She also spontaneously reported experiencing a shorter latency to sleep onset, a more restful sleep, as indicated by a reduction of her normal physical reorientation in bed through a night, and a more rapid awakening in the morning. None of the latter changes could be documented objectively, but they were particularly interesting in terms of the similar, quantified findings obtained with SMR in the cat. (Sterman & Friar, 1972, p. 91)

Sterman did not only develop theories about sleep and epilepsy. His observation of behavioral changes in the secretary and the cats is important to understand how contemporary brain device users work on themselves. Where Berger visualized the 'writings' of the human brain, Walter turned this brain into a living or performative brain that could act upon the self, Kamiya introduced a self that could act upon the brain, and Sterman elaborated on this controlling self by claiming that the subject's personality (as well as the sleeping patterns and epileptic fits) could be altered through conditioning. He described his method in terms of 'voluntary control', 'voluntary therapy', and stated: 'The method of biofeedback requires that the subject assumes personal responsibility for any beneficial effect to be had and provides the basis for a new level of self-awareness' (Sterman, 1981). With this, he makes clear that the self can and should control its brain and actions.

In media interviews and articles, on the other hand, Sterman makes several comments in which the self is completely controlled by the brain. In a popular book on neurofeedback, for example, he is quoted as saying:

> I can tell from an EEG whether someone's paying attention, and if they are, if they are paying attention to me or to what they did last night. You can tell whether someone is mildly retarded from an EEG. Or whether someone is hyperaroused and can't relax. (…) Everything depends on the topographical distribution. (Robbins, 2000, p. 34)

Together with a colleague, Sterman nowadays runs a company for 'evidence-based neurotherapy', and their website contains phrases like 'bad brain habits' and 'neuromodulations for each disorder'.[10] They visualize these disorders as colorful brain states in a picture.[11] Under this picture they enumerate the disorders they can modulate, varying from 'brain injury' and 'emotional disturbances', to 'efficacy', 'social functions', and 'love'.[12] That is, Sterman's ideas about the brain and the self are actually rather complex since they include two actors: the brain works upon the self, and the self works upon the brain.

This struggle between the brain and the self becomes manifest in 2001, when Sterman is cited in several American newspaper articles because he testified for the defense of a convicted murderer. The defendant, Terry Clark, was sentenced to death for the murder of a nine-year-old girl in New Mexico in 1986. Based on a brain scan made by a colleague, Sterman declared: 'It's a disability, not a bad person' and 'I don't think you want someone with his frontal lobe disturbance out in society' (Bresenham, 2001; Herrera, 2001a). However, the defendant himself claimed that 'these reports about me having brain damage are false' and 'I have a personal, moral and social obligation to take responsibility for what I did' (Herrera, 2001b; 'New Mexico: Death-row inmate says he's not brain-damaged' 2001).[13] A few months after, Clark, who had voluntarily stopped his appeals procedure, was executed with a lethal injection. That is, in contrast to Walter, who had successfully defended a murderer by showing with an EEG that the defendant had had an epileptic seizure, Sterman encountered a defendant who resisted his brain diagnosis because he wanted to take responsibility for his own actions.

Perhaps this anecdote is telling for Sterman's somewhat ambiguous position regarding the steering brain and the modulating self. Neuromodulation is a 'voluntary therapy' that provides self-awareness, and requires that the subject take responsibility. On the other hand, Sterman (and/or his colleague) categorizes phenomena like murder, social functioning, and love under brain-based behavior. This appears to be contradictory, and can

---

[10] www.skiltopo.com/index.php (accessed on 14-06-12).

[11] See http://www.skiltopo.com/ClinicalResearch/summaries.php (accessed on 28-09-2015).

[12] An article of Sterman's colleague David Kaiser clarifies: 'Love is the primary source of neuroplasticity', and neuromodulation is a form of 'guided neuroplasticity'.

[13] Clark does not directly respond to Sterman with this citation. Several other experts claimed that Clark had brain damage.

perhaps best be understood by returning to Walter who claimed that his robotic tortoises were also capable of human and social behavior, such as self-recognition and acting out of free will. That is to say, Walter and Sterman both invented a determining brain, but combined their ideas about brain brothers and brain habits, with a certain form of free will.

The Mind–Body Web

Sterman's work is cited more often, but Kamiya's experiments were well received in spiritual circles. This spiritual connotation might explain why EEG biofeedback largely disappeared from the psychological stage.[14] However, over the past few years, the use of EEG biofeedback has made its comeback in the form of neurofeedback, and nowadays the therapy is increasingly offered by private clinics. Practitioners working in these clinics do not only promote and explain their therapies, they also spread the message that people can work on their selves by training their brains, and make their clients familiar with their (problematic) brainwaves, brain maps, and potential personality changes.

The motivation to promote this message is often encouraged by personal experiences with neurofeedback. The majority of practitioners started as clients, or as users in another sense (self-experimenter, test subject), and became so enthusiastic about the therapy that they opened their own clinics. Others were initially motivated by the wish to cure their children, for example, from ADHD or learning difficulties, and those who started with a strictly scientific or therapeutic drive mostly started to use neurofeedback on themselves, later on. That is to say, just like Berger, Walter, Kamiya, and Sterman were prompted by private beliefs and experiences—telepathy, cybernetics, the relevance of subjective experiences,

---

[14] EEG biofeedback never entirely disappeared. In the USA, several other figures were important in the establishment (or continuation) of neurofeedback. An assistant of Sterman, Margaret Ayers, noticed that after doing neurofeedback, 'these epileptic individuals were happier, smiling, they were talking about things' (Robbins, 2000). She started one of the first neurofeedback clinics in the USA. One of her clients was the son of 'The Othmers' who became prominent neurofeedback promoters after their son had benefitted from Ayer's training in the 1980s. The American psychologist Joel Lubar is important because of his EEG biofeedback research on hyperactive children (Lubar & Shouse, 1976).

a more outgoing secretary; the motivations of contemporary practitioners are also often personal: most contemporary practitioners have their experiences with recovered selves, friends, or family members.

Another aspect that can be traced in the stories of neurofeedback 'pioneers' as well as in those of the current promoters is the ambivalent relation between the brain and the self. Berger tried to grasp the psyche by visualizing brain activity, but he became more and more frustrated about the complex 'interaction' between the two. He tried to deny his own struggling psyche by pretending heart problems, and ended this struggle by taking his life. Walter tried to liberate the controllable brain from the intangible self with his cybernetic theories, but on several occasions his brain ideas were endangered, and he increasingly started to worry about the consequences of his theories. Kamiya intended to pay attention to subjective experiences (feelings, hopes) by teaching people to train their own brain state, which resulted in a trend of people trying to change their own brain state to reach a more spiritual self. And in Sterman's work the brain is controlled by the self, but the self is also controlled by the brain, and this ambiguity is brought to a head by the murderer who denies brain damage and wants to take responsibility himself. In all these cases the brain and the self do not simply coincide, but seem to struggle for control.

This complicated relationship between the brain and the self can also be retrieved in the acts and statements of contemporary practitioners. To become neurofeedback experts, they have to transform their thoughts and explanations from a psychological into a brain idiom. As formulated in a neurofeedback course, one of the first things they have to do to become neurofeedback specialists is to 'learn to focus on symptoms as signs of brain physiology, not psychology'.[15] The supervisor explains this with an example—when a client complains that he has problems in falling asleep because there is 'lots more going on in his life', the practitioner should be aware, and wonder: 'Do you talk about it—or change the NF training?'. And the advice is, 'Consider that neurofeedback played a role, till you rule it out.' That is, this supervisor emphasizes that complaints in general or during the training might be a result of (changing) brainwaves. When one of the participants jokes during the course: 'Now clients

---

[15] Quote from a sheet presented during a neurofeedback course for novice practitioners.

can say: "It wasn't my fault, it was my brain"', one of the supervisors firmly answers: 'And that is exactly right!' That is, not only in this course, but among neurofeedback practitioners in general, the idea that 'the mind is the brain' is an important statement.

However, this principle 'the mind is the brain' is also most easily abandoned by neurofeedback specialists; for example, when they state that 'poor parenting does not cause ADHD, but can make it worse' (3) or by stressing that someone should not be 'entirely in the hands of his central nervous system' (7). When being directly confronted with questions concerning the mind–body relation practitioners occasionally avoid the question. They say, for example, 'I cannot think about what that really means' (5), or 'You cannot separate the mind and the brain. There is a link. So if you work on the brain you can change the mind, don't you?' (2). Another practitioner becomes confused when he tries to explain the difference between brain behavior and cognitive behavior:

> It is not only that you are angry because of your brain behavior, your brain behavior will also… Well, yes, what am I saying…? Of course cognitive behavior is brain behavior too. Yes, now it becomes really… Before you notice you are in the neuro-philosophical corner, which is very interesting, but it is not about feedback. (7)

It is not surprising that practitioners use this kind of confusing language when confronted with mind–body issues. As I earlier demonstrated even 'thorough-going materialists' like Walter can get caught in the web of mind–body interactions. Moreover, as the philosopher Ian Hacking concluded: 'Neuroscience is not so monistic as it so confidently asserts' (Hacking, 2007, p. 101) since contemporary neuroscientists keep dividing people into beings with mind, brain, and body qualities, like emotions, thoughts, and sensations (Hacking, 2004, 2005). A good example comes from the popular Dutch neuroscientist Dick Swaab who claims that 'We are our brains', but also that the brain produces the mind, just like the kidney produces urine (Swaab, 2014; Heijden van der, 2011).[16] The second statement already adds a mind to this brain creature.

---

[16] This expression can be compared with Walter's 'velocity versus the engine', with the difference that urine is a substance while velocity is a capacity. This formulation suggests that the mind is a material entity and separated from the brain.

Aside from these puzzling statements, neurofeedback practitioners also use language that comes close to the terminology of Walter, when he divided people into personalities with regard to their brain types. One of the supervisors during the neurofeedback course, for example, clarifies the difference between two levels of dopamine by stating: 'You can see low levels walking into your door [acts lethargic] and you can see high levels [acts hyperactive].' During the same course, several other connections between someone's brain and personality are made and put forward in stereotypes. Alcoholics are called 'alphaholics'[17]; brainwaves are personalized by calling them nice, beautiful, or very reactive; and when one of the supervisors uses a metaphor, his colleague explains: 'He is very good at metaphors, he has a good parietal lobe.'[18]

Other practitioners explain how they increasingly learned to connect brain activity to behavior during their careers. Someone says, 'You started looking at people and thought: "This person needs somewhat more beta", or "that person needs SMR"' (7). Another practitioner explains how he recognizes people with high frequencies in their brains: 'What you often see with people with high beta, those who really bite-the-mind you know, that everything spins around [in their head] and they can't stop it. You notice they breath very high. (...) and you notice they sit like this [hunches his shoulders, shrinks his body]' (6). His colleague speaks of persons with low alpha waves as 'low-voltage persons', which comes close to Walter's M (minus) alpha type.

Another recurring theme that started with Berger's interest in telepathy and can be traced in the work of all cited scientists is the surprising connection between a materialistic point of view and a spiritual way of thinking. Several neurofeedback users also practice yoga or meditation, some practitioners use neurofeedback to meditate or to hypnotize themselves, and others learnt about the therapy by reading spiritual magazines or books. People who try to work on their brains to enhance themselves often use reductionist language (the mind is the brain), but also easily connect this way of thinking with a holistic view. That is to say, believing that the self is a total functioning organism (holism) or reducing it to

---

[17] Alpha waves are related to a calm state, and drinking alcohol appears to evoke alpha waves.
[18] See the introduction of Chap. 6.

the brain (monism) practically have the same consequences: we can start working on our brainwaves, or neurotransmitters, or brain spots, with the purpose of enhancing our selves.

Conclusion

Berger tried to grasp psychical energy and in his devotion he created a brain that could give signals on paper. Walter claimed not to be interested in concepts like the mind and introduced a performative brain that appeared to be able to control the self. Kamiya constituted a brain that could be trained by the self in order to help this self becoming more spiritual. And Sterman designed a brain that could control, and could be controlled by, the self. In this development, several brain-related entities (alpha, beta, theta, SMR) have been distinguished. Connections and entanglements between human and machine emerged, mind–body problems were enlarged, materialistic philosophies were combined with spiritual ideas, and struggles between brains and selves appeared. When comparing these findings with the reports of contemporary neurofeedback practitioners, some analogies can be distinguished. Neurofeedback experts are not only often inspired by private experiences and spiritual beliefs, in their explanations the brain becomes a steering actor that sometimes completely substitutes the self ('low levels [of dopamine] walking into your door'), while they simultaneously refer to another actor like the mind or the self (someone should not be 'entirely in the hands of his central nervous system').[19] That is to say, this chapter demonstrates that working on the brain to understand or improve the self does not simply reduce the self to the brain, but appears to evoke an interaction between these entities. In Chaps. 4 and 6 I study how neurofeedback users and practitioners experience and manage this 'ontological difficulty', as I called it before. Although neurofeedback users are generally not philosophers bothered by ontological difficulties (see also Chap. 7), they do show that working on the brain to improve the self does not reduce the self to the brain but extends the self. The next chapter gives an

---

[19] See also Chaps. 4 and 6 for more examples.

account of contemporary neurofeedback users who constitute themselves as a blend of psychological, physiological, mechanical, and spiritual entities. And although such amalgamations of brains and selves, humans and machines, and material and spiritual philosophies could be dismissed as being confusions of lay people confronted with a new ('modern') scientific way of thinking, the current chapter demonstrated that this could better be interpreted as a result of the (historical) quest to grasp the self with a brain device.

References

Adrian, E. D., & Matthews, B. H. C. (1934). The Berger rhythm: Potential changes from the occipital lobes in man. *Brain, 57*(4), 355–385.

Amen, D. G. (1998). *Change your brain, change your life: The revolutionary, scientifically proven program for mastering your moods, conquering your anxieties and obsessions, and taming your temper.* New York: Times Books.

Berger, H. (1910). *Untersuchungen über die Temperatur des Gehirns.* Jena: Fischer.

Berger, H. (1940). *Psyche.* Jena: Fischer.

Berger, H. (1969). *Hans Berger. On the electroencephalogram of man. The fourteen original reports on the human electroencephalogram.* (P. Gloor, Trans.). Amsterdam: Elsevier.

Boening, H. (1941). Professor Hans Berger-Jena. *Archiv für Psychiatrie und Nervenkrankheiten, 114*(1), 17–24. doi:10.1007/BF02047252.

Borck, C. (2001). Electricity as a medium of psychic life: Electrotechnological adventures into psychodiagnosis in Weimar Germany. *Science in Context, 14*(4), 565–590.

Borck, C. (2005a). *Hirnströme. Eine Kulturgeschichte der Elektroenzephalographie.* Göttingen: Wallstein Verlag.

Borck, C. (2005b). Writing brains: Tracing the psyche with the graphical method. *History of Psychology, 8*(1), 79–94.

Brazier, M. A. B. (1961). *A history of the electrical activity of the brain: The first half-century.* London: Pitman.

Bresenham, J. (2001). Testimony indicates murderer disabled. *Amarillo Globe-News.* Retrieved from http://amarillo.com/stories/2001/07/31/new_indicates.shtml

Budzynski, T. H. (1999). From EEG to neurofeedback. In J. R. Evans & A. Abarbanel (Eds.), *Introduction to quantitative EEG and neurofeedback* (pp. 65–79). San Diego, CA: Academic.

Churchland, P. S. (2002). *Brain-wise: Studies in neurophilosophy*. Cambridge, MA: MIT Press.

Craver, C. F. (2009). *Explaining the brain*. Oxford, USA: Oxford University Press.

Damasio, A. (2012). *Self comes to mind: Constructing the conscious brain*. New York: Random House Incorporated.

de Spinoza, B. (2006). *The ethics*. Teddington, Middlesex: Echo Library.

Demos, J. N. (2005). *Getting started with neurofeedback*. New York/London: W.W. Norton.

Dennett, D. C. (1991). *Consciousness explained*. Boston: Little, Brown and Company.

Doidge, N. (2007). *The brain that changes itself: Stories of personal triumph*. London (etc): Penguin.

Electronic patterns of the brain. (1956). *Life, 40*(15), 89–90.

Geiger, J. (2003). *Chapel of extreme experience. A short history of stroboscopic light and the dream machine*. Brooklyn, NY: Soft Skull Press.

Gibbs, F. A. (1941). Prof. Dr. Hans Berger 1873–1941. *Archives of Neurology and Psychiatry, 46*(3), 514–516. doi:10.1001/archneurpsyc.1941.0228021 0140013.

Ginzberg, R. (1949). Three years with Hans Berger a contribution to his biography. *Journal of the History of Medicine and Allied Sciences, 4*(4), 361–371.

Gloor, P. (1969). Hans Berger and the discovery of the electroencephalogram. In H. Berger (Ed.), *Hans Berger. On the electroencephalogram of man* (pp. 1–25). Amsterdam/London/New York: Elsevier.

Gloor, P. (1994). Berger lecture *. Is Berger's dream coming true? *Electroencephalography and Clinical Neurophysiology, 90*, 253–266.

Hacking, I. (2004). Minding the brain. *New York Review of Books, 51*(11), 32–36.

Hacking, I. (2005). The Cartesian vision fulfilled: Analogue bodies and digital minds. *Interdisciplinary Science Reviews, 30*(2), 153–166.

Hacking, I. (2007). Our Neo-Cartesian bodies in parts. *Critical Inquiry, 34*, 78–105.

Harrington, A. (1999). *Reenchanted science: Holism in German culture from Wilhelm II to Hitler*. Princeton, NJ: Princeton University Press.

Hayward, R. (2001). The tortoise and the love-machine: Grey Walter and the politics of electroencephalography. *Science in Context, 14*(4), 615–641.

Herrera, P. (2001a, July 31). He is a bad person. *Santa Fe New Mexican*.

Herrera, P. (2001b, August 10). Judge says Terry Clark competent; sets Nov. 6 execution. *The Associated Press State & Local Wire*.

Holland, O. (2003). Exploration and high adventure: The legacy of Grey Walter. *Philosophical Transactions of the Royal Society of London. Series A: Mathematical, Physical and Engineering Sciences, 361*(1811), 2085–2121.

Jung, R. (1963). Hans Berger und die Entdeckung des EEG nach seinen Tagebüchern und Protokollen. In R. Werner (Ed.), *Jenenser EEG_Symposion. 30 Jahre Elektroenzephalographie. 17–19. Oktober 1959* (pp. 20–53). Berlin: VEB Verlag Volk und Gesundheit.

Jung, R., & Berger, W. (1979). Fünzig Jahre EEG. Hans Berger Entdeckung des Elektrenkephalogramms und seine ersten Befunde 1924–1931. *Archiv für Psychiatrie und Nervenkrankheiten, 227*, 279–300.

Kamiya, J. (1968). Conscious control of brain waves. *Psychology Today, 1*(11), 56–60.

Kamiya, J. (1969). Operant control of the EEG alpha rhythm and some of its reported effects on consciousness. In C. T. Tart (Ed.), *Altered states of consciousness. A book of readings* (pp. 507–517). New York/London/Sydney, NSW/Toronto, ON: Wiley.

Kamiya, J. (1971). Operant control of the EEG alpha rhythm and some of its reported effects on consciousness. In J. Kamiya, T. X. Barber, L. V. DiCara, N. V. Miller, D. Shapiro, & J. Stoyva (Eds.), *Biofeedback and self-control: An Aldine reader on the regulation of bodily processes and consciousness* (pp. 280–290). Chicago/New York: Aldline Atherton.

Kamiya, J. (2011). The first communications about operant conditioning of the EEG. *Journal of Neurotherapy, 15*(1), 65–73. doi:10.1080/10874208.2011.5 45764.

LeDoux, J. E. (2003). *Synaptic self: How our brains become who we are.* New York: Penguin.

Lubar, J. F., & Shouse, M. N. (1976). EEG and behavioral changes in a hyper-kinetic child concurrent with training of the sensorimotor rhythm (SMR). *Biofeedback and Self-Regulation, 1*(3), 293–306.

Luce, G., & Peper, E. (1971, September 12). Mind over body, mind over mind. *The New York Times Magazine*, pp. 34–35.

Millett, D. (2001). Hans Berger. From psychic energy to the EEG. *Perspectives in Biology and Medicine, 44*(4), 522–542.

New Mexico: Death-row inmate says he's not brain-damaged. (2001, May 3). Albuquerque: The Associated Press. Retrieved from http://amarillo.com/stories/2001/05/03/new_inmate.shtml#.VsM7s5dM6ZY

Niedermeyer, E., & Lopes da Silva, F. H. (2005). *Electroencephalography: Basic principles, clinical applications, and related fields.* Philadelphia: Lippincott Williams & Wilkins.

Pickering, A. (2008). *Brains, selves and spirituality in the history of cybernetics.* Workshop paper, Arizona State University. Retrieved from http://hdl.handle.net/10036/81576

Pickering, A. (2010). *The cybernetic brain. Sketches of another future.* Chicago/London: The University of Chicago Press.

Robbins, J. (2000). *A symphony in the brain: The evolution of the new brain wave biofeedback*. New York: Atlantic Monthly.

Sourkes, T. L. (2006). On the energy cost of mental effort. *Journal of the History of the Neurosciences, 15*, 31–47.

Spear, J. H. (2004). Cumulative change in scientific production: Research technologies and the structuring of new knowledge. *Perspectives on Science, 12*(1), 55–85. doi:10.1162/106361404773843346.

Sterman, M. B. (1981). EEG biofeedback: Physiological behavior modification. *Neuroscience and Biobehavioral Reviews, 5*, 405–412.

Sterman, M. B., & Egner, T. (2006). Foundation and practice of neurofeedback for the treatment of epilepsy. *Applied Psychophysiology and Biofeedback, 31*(1), 21–35.

Sterman, M. B., & Friar, L. (1972). Clinical Note. Suppression of seizures in an epileptic following sensorimotor EEG feedback training. *Electroencephalography and Clinical Neurophysiology, 33*, 89–95.

Sterman, M. B., LoPresti, R. W., & Fairchild, M. D. (2010). Electroencephalographic and behavioral studies of monomethyl hydrazine toxicity in the cat. *Journal of Neurotherapy, 14*(4), 293–300. doi:10.1080/10874208.2010.523367.

Sterman, M. B., & Wyrwicka, W. (1967). EEG correlates of sleep: Evidence for separate forebrain substrates. *Brain Research, 6*, 143–163.

Swaab, D. F. (2014). *We are our brains: A neurobiography of the brain, from the womb to Alzheimer's*. New York: Spiegel & Grau.

van der Heijden, M. (2011). Een goede tegenwerping heb ik nog niet gehoord. *NRC handelsblad*.

Velmans, M. (2000). *Understanding consciousness*. London: Routledge.

Walter, W. G. (1952). Patterns in your head. *Discovery, 13*, 56–62.

Walter, W. G. (1953a). *The living brain*. London: G. Duckworth.

Walter, W. G. (1953b). The electroencephalographic development of children. In J. M. Tanner & B. Inhelder (Eds.), *Discussions on child development* (Vol. 1, pp. 132–140). London: Tavistock Publications.

Walter, W. G. (1956). Comments on professor Piaget's paper. In J. M. Tanner & B. Inheleder (Eds.), *Discussions on child development* (Vol. 4, pp. 53–60). London: Tavistock Publications.

Walter, W. G. (1957). *The living brain*. London: Gerald Duckworth & Co. Ltd.

Walter, W. G. (1960). *Accomplishments of an artefact*. Three short talks accompanied by fourteen slides, illustrating the learning process in M. Speculatix; M. Docilis and a human subject connected to an EEG monitor.

Walter, W. G. (1968a). The social organ. *Impact of Science on Society, 18*(3), 179–186.

Walter, W. G. (1968b). The past and prospect of EEG. In G. Alemá, M. Gozanno, & R. Vizioli (Eds.), *Brain and mind problems* (pp. 755–764). Rome: "Il Pensiero Scientifico" Publishers.

Walter, W. G. (1968c). Electronic devices in psychiatry, a commentary. In N. S. Kline & E. M. Laska (Eds.), *Computers and electric devices in psychiatry* (pp. 135–140). New York: Grune & Stratton.

Walter, W. G. (1969a). *Observations on man, his frame, his duty and his expectations.* Cambridge, MA: University Press.

Walter, W. G. (1969b). Neurocybernetics (communication and control in the living brain). In J. Rose (Ed.), *Survey of cybernetics. A tribute to Dr. Norbert Wiener* (pp. 93–108). London: Ilife Books.

Walter, W. G. (1971). The future of clinical neurophysiology. In A. Rémond (Ed.), *Handbook of electroencephalography and clinical neurophysiology* (pp. 1a: 39–50). Amsterdam: Elsevier.

Walter, W. G. (1972a). My miracle. *Theoria to Theory, 6*, 38–50.

Zimbardo, P. G., & Maslach, C. (1977). *Psychology for our times: Readings.* Glenview, Ill: Scott, Foresman.

4

Taking Care of One's Brain

'Shocking, isn't it?' says the neurofeedback practitioner and he starts to laugh. 'What?' I respond, while we are watching the real-time scribbles that are produced by my brain. 'You have much muscle tension, but that is not so uncommon', he answers, 'Please, try to relax, and stop blinking so frequently.' It is surprisingly hard to sit still, stare, and suppress my blinking for five minutes, and apparently I am not very good at it either, since five minutes later the practitioner repeats that my recordings show many artifacts. To collect all information that is needed to produce my quantitative electroencephalogram (qEEG) he asks me to close my eyes again, and sit still and relax for another five minutes. When I am done with this second exercise, the practitioner asks me what I want to train. I am somewhat puzzled by his question and have no idea, but the practitioner has, and suggests that I can use some alpha to make me feel more relaxed. Again, I can close my eyes for 20 minutes, but this time while listening to a piece of spiritual music, rolling waves, beeps, and once in a while a screaming seagull. I feel tense: the music is annoying, the seagull makes me laugh, and I hear the practitioner typing, and walking around the room.

When the alpha training is over, the practitioner asks me how I feel, and if I noticed anything. I don't know what to answer and ask him what he thinks that I should have noticed. He wants to know if I had control.

J. Brenninkmeijer, *Neurotechnologies of the Self,*
DOI 10.1057/978-1-137-53386-9_4

I demur, since I really cannot understand how and of what (my brain, the computer?) I should get control. The practitioner reassures me that it is not necessary to feel control, since the brain picks it up anyway, but also reminds me that I should interpret the beeps as rewarding, since my brain produces the right frequencies whenever these beeps occur. He proposes to do another session, a beta training this time, which will make me feel alert again. Once more I may listen for 20 minutes to some spiritual music, accompanied by beeps and a roaring waterfall this time. I try, but I cannot figure out how to concentrate on, let alone control, the beeping of the beeps. I feel like I have wound up in a meditational exercise, in which I do not master the techniques.

A few days later, the practitioner sends me an e-mail with the text: 'Here comes your qEEG.' Attached are two documents with a total of 100 pages of green, yellow-, red-, and blue-colored heads, but without any explanation of the meanings. I am puzzled, and feel uncomfortable. What does this person know about me that I don't?

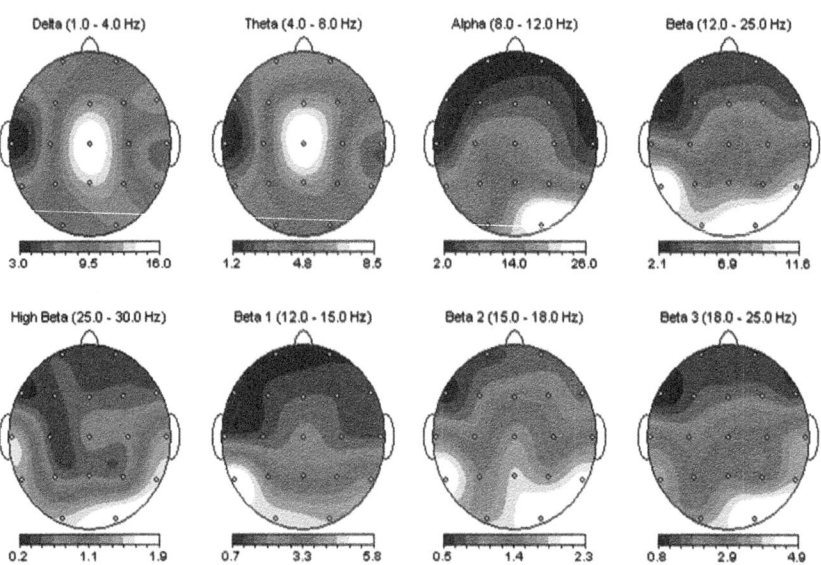

FFT Absolute Power (uV Sq)

| Delta (1.0 - 4.0 Hz) | Theta (4.0 - 8.0 Hz) | Alpha (8.0 - 12.0 Hz) | Beta (12.0 - 25.0 Hz) |

| 3.0   9.5   16.0 | 1.2   4.8   8.5 | 2.0   14.0   26.0 | 2.1   6.9   11.6 |

| High Beta (25.0 - 30.0 Hz) | Beta 1 (12.0 - 15.0 Hz) | Beta 2 (15.0 - 18.0 Hz) | Beta 3 (18.0 - 25.0 Hz) |

| 0.2   1.1   1.9 | 0.7   3.3   5.8 | 0.5   1.4   2.3 | 0.8   2.9   4.9 |

Fig. 4.1   One page of my qEEG (Used with permission of Roland Verment)

Most people go to a neurofeedback clinic for a specific reason. They are, for example, diagnosed with attention deficit disorder (ADD), feel like having a burnout, want to improve their performances, or have problems with sleeping. My own experiences, however, were not related to a specific complaint. I went to do neurofeedback because I wanted to experience what other users' experience. Hence, during the neurofeedback I was not a cooperative client who wanted to change her brain, but a researcher who observed the situation, and sometimes paid more attention to the practitioner than to the training. When the practitioner was tuning the frequencies of the neurofeedback program, I was observing him. When he walked in and out of the room, I was struck by the fact that he mostly watched his computer screen instead of his client. (His client was on his computer screen!) At some points, my trainings became really fascinating experiences, for example, when the practitioner showed the documentary 'A transcendent man' about the futurist Ray Kurzweil: I was sitting in a comfortable chair, my head connected (and fixated) to a computer by several EEG wires, watching a movie that enlarged and reduced with my brain activity, while (the in and out zooming) Ray Kurzweil insisted that 'We are all becoming cyborgs.' At another moment, and with another practitioner, however, I was listening to a spiritual reading of Eckhart Tolle who talked about stillness and being yourself.

My experiences with Kurzweil and Tolle are illustrative of the amalgamation of technical and spiritual ideas that is so characteristic of neurofeedback, but also of the tension between the two; becoming a cyborg, one might intuitively say, opposes stillness and being yourself. However, there are also some analogies between Kurzweil and Tolle since both 'teachers' encourage people to work on themselves, to become a better (enhanced or more spiritual) being. Although the practitioners could have chosen different movies to show—varying from cartoons, science fiction, thrillers, to comics—their choices also reveal some of the ideas and purposes behind neurofeedback.

Another characteristic that became obvious by doing neurofeedback is that the relationship between the client and the practitioner mainly goes via the computer. The client performs the training, for example, by relaxing, concentrating, and trying to retrieve control. In the meantime, the practitioner watches the client's efforts on the computer screen and tunes some frequencies, and only pays attention to the client before and after the training when he asks how the client feels. The puzzling qEEG and the shocking brainwaves are illustrative of the complicated (indirect) relationship between the client and the practitioner. The practitioner and the client are both watching the same real-time information (quantified and visualized by a computer), but the client is completely dependent on the information the practitioner gives.

As I experienced myself by doing neurofeedback, this information can have quite some impact. Watching my own brainwaves was a rather strange and somewhat awkward experience, especially since the practitioner mentioned the word 'shocking'. During the trainings, I mostly felt uncomfortable because the question of what the practitioner and his computer were actually doing with my brain—and which effects this could have—kept on haunting me. I also felt annoyed with the scant information I received about my fully colored brain map and angled for more information, whenever I had the chance. Questions like 'If I had control' during the training and 'If I noticed something' since the training last week affected me too. Not in the sense that I felt control, or noticed any changes, but in the sense that I did not want to feel control (me, having control over my brain?), nor any changes (myself, being changed by a computer?). Actively doing neurofeedback did not only make me more aware of my brain than reading or hearing about it; to a certain extent I seemed to feel endangered in my sense of self.

The circumstance that doing neurofeedback slightly influenced my thinking about myself in a period of only five sessions emphasizes the question which effect neurofeedback has on people who do 20–60 sessions with the purpose of changing themselves. In this chapter I will try to find out how people who use neurofeedback to treat themselves for psychiatric disorders, not functioning optimally, or other complaints or purposes reconceptualize their selves[1] and their problems. My interest to study neurofeedback as a technology of the self was evoked among

---

[1] People who want to change *themselves* can use techniques that change *their selves*: the difference between changing themselves (feeling better) and changing their selves (subjectivity, identity) is confusing but will become clear in the course of this book.

others, by the work of the sociologist of science Andrew Pickering who studied cybernetic practices and writes about the flicker experiments of Walter, discussed in the previous chapters:

> I can't help thinking of Michel Foucault's idea of *technologies of the self*. In Foucault's own work, these are technologies that produce a distinctly human, self-controlled self—the kind of self that sets us apart from animals and things. Flicker, then, is a different kind of nonmodern, non-Cartesian technology of self—a technology for *losing control* and going to unintended places, for *experiment* in a performative sense. (Pickering, 2008, p. 5)

According to Pickering, Walter's experiments in which he stimulated the brain with light pulses produced a non-modern[2] and non-Cartesian self; that is, a self in which the distinctions between subject and object, or between mental and material, are blurred. Pickering's and Foucault's work made me wonder what kind of self is constituted by doing neurofeedback—a technology not used for losing, but for retrieving control.

Technologies of the Self

In his *History of Sexuality* (Foucault, 1990a, 1990b, 1992) Foucault described how people since antiquity have constituted their identity by using various 'techniques' such as reading manuscripts, keeping diaries, making confessions, listening to teachers, or saying prayers. Techniques of the self:

> permit individuals to effect by their own means or with the help of others a certain number of operations on their own bodies and souls, thoughts, conduct and way of being, so as to transform themselves in order to attain a certain state of happiness, purity, wisdom, perfection, or immortality. (Foucault, 1988)

For Foucault, working upon oneself—an exercise of the self on the self—with the purpose of developing and transforming oneself, is an

---

[2] With 'non-modern' Pickering refers to Latour (1993) who states that non-moderns do not make a clear distinction between subjects and objects, while modern people do.

attempt to attain a certain mode of being. He describes how this constituting of oneself is related to three axes, namely, knowledge, power, and ethics: we constitute ourselves as 'objects of knowledge', as 'subjects acting on others', and as 'moral agents'. In other words, if we take proper care of ourselves—if we know what we are, what we are capable of, reasonably hope for, should fear, how we should act on others, and so on—we can constitute ourselves as moral agents (Foucault, 1997a). Foucault's critical or historical 'ontology of ourselves' does not refer to the assumed biological or neurological processes we are made up of, but to a certain way of being and knowing oneself. Users of brain devices, however, are specifically confronted with a biological or neurological way of knowing themselves. This makes it interesting to find out what kind of self people who try to change their brain constitute. What is the mode of being that brain device users strive for? What do they hope and fear? What is the knowledge through which they constitute themselves, and what kind of self is the result?

Constituting a certain mode of being is related to what Foucault calls the 'ethical substance', that is, the aspect of the self that is concerned with moral conduct. For Foucault, the aspect of the self which is supposed to make us better beings changes in times and cultures. Roughly stated, the 'ethical substance' was desire for the Christians, intentions for Kant, and feelings in the 20th century (Foucault, 1997b, p. 263). Christians worked on themselves, for example, by using their sexual desires for reproduction, and not for masturbation. According to Kant, we had to work on our intentions to conform oneself to universal rules to become a good, universal subject. The need for 'coming out' for homosexuals (Foucault, 1997c, p. 139) is an example that illustrates that feelings were especially relevant for ethical judgment in the 20th century.

In the *History of Sexuality* Foucault demonstrated that the injunction 'Take care of yourself' was a very important precept for the Greeks, while 'Know yourself' was the most central precept for the Christians, and he explains that these different forms of care mean different forms (as well as technologies) of self.[3] He also refers to the difference between 'knowing

---

[3] Both principles are from antiquity, but according to Foucault the meaning of 'know yourself' changed and became more important during Christianity (Foucault, 1984, 1988).

yourself' during Christianity—which relied heavily on the technique of verbalization in which disclosure is combined with renunciation—and the modern technique of verbalization from the human sciences that includes the disclosure, but not the renunciation, of the self. This new technique of verbalization, Foucault writes, positively constitutes a new self (Foucault, 1988, p. 49, see also 1984).

According to Foucault, technologies of the self exist in every culture, but the principles, concepts, and cultures of self can be completely different. He gives the examples of an ancient 'culture' of self, and what he calls a Californian 'cult' of self.

> In antiquity, this work on the self with its attendant austerity is not imposed on the individual by means of civil law or religious obligation, but is a choice about existence made by the individual. People decide for themselves whether or not to care for themselves. (…) In the Californian cult of the self, one is supposed to discover one's true self, to separate it from that which might obscure or alienate it, to decipher its truth thanks to psychological or psychoanalytic science, which is supposed to be able to tell you what your true self is. Therefore, not only do I not identify this ancient culture of the self with what you might call Californian cult of the self, I think they are diametrically opposed. (Foucault, 1997b, p. 271)

Foucault's work is mainly historical, and in his explanation of technologies of the self (which is only part of his work—see intermezzo) he especially describes ancient and Christian techniques. He discusses 'modern' self practices just briefly by referring to the human sciences as distributors of techniques of verbalization by which people constitute themselves. The sociologist Nikolas Rose, among others, took up Foucault's work and demonstrated that verbalization indeed became a very important technique of the human or social sciences, especially of the 'psy disciplines', a word used by Rose to refer to all disciplines with names that start with 'psy', such as psychology, psychiatry, or psychotherapy (Rose, 1998). For several decades, psy disciplines have described 'the human being' by using questionnaires, interviews, psychological tests, and therapeutic methods, and these practices had an enormous influence on people's personal self. Techniques like confessing your sins to a priest and revealing your thoughts to a therapist or questionnaire are techniques of self because people use them to constitute and reveal the truth about their selves. However, by collecting and combining all those personal truths, human scientists have a very powerful technology in their hands with which they can create the 'general' truth of the 'human being'. In this sense, the technique of verbalization,[4] as used by human scientists, works like a double-edged sword. People constitute their own truth, but by doing this they are simultaneously mirrored in the truth of the general human being, which reflects upon their own truth. Hence, following Foucault, some scholars claim that social scientists not only describe but also *inscribe* people (Hacking, 1999a, 2006).

However, new insights and premises about the brain have drastically changed the psy disciplines, and brain scans, EEG devices, and technical or pharmacological brain manipulators have become more and more important as technologies to inscribe people (Rose, 2007). People increasingly learn that their daily life problems are brain problems and from this point of view it is not surprising that some of them start to experiment with manipulating their brain. Probably, working upon one's feelings, by means of exposing or showing these with techniques of verbalization, is not the most common way people work upon themselves anymore these days. For a decade or two, sources, like newspaper articles, self-help books, and scientific publications, give the impression that the brain is increasingly seen as the part of oneself that defines moral conduct (Amen, 1998; Churchland, 2011; Doidge, 2007). With this change in ethical substance, the 'mode of subjectivation'—that is, the way we style ourselves—also changes and makes the use of brain therapies more relevant. Especially for people who use a brain device to cure or enhance themselves, the ethical substance appears to be their brain, or inside of their brain. However, this also means that the part of oneself that those people see as responsible for their behavior changed from something psychological (feelings, desires, intentions) to something biological (neurons, brainwaves). Especially in the case of neurofeedback this change of substance seems to become translated in a literal enactment of the mind/body problem, because people stare at their material waves in order to transform their immaterial feelings.

---

[4] Besides verbalization the human sciences make use of various other techniques such as diagnostic handbooks, statistics, and non-verbal therapies (see, for example, Rose, 1998).

How do people handle this transformation? How do they fit their brain-waves into their knowledge of their selves, their problems, their hopes and fears, and their relationships with others?

Restore the Self by Restoring the Brain

People use neurofeedback for various purposes, from achieving peak performances or improving meditation skills, to treating mental or physical disorders. In interviews and questionnaires, the reasons people gave for doing neurofeedback were often stress-related. Other motivations were being diagnosed (or having diagnosed oneself) with ADHD, feeling depressed or anxious, having problems with concentration, or not functioning 'normally'.[5] In general, my respondents hoped to get rid of their problems, to start functioning normally, or to improve their quality of life. About 75 percent of my respondents—and this corresponds with the claim practitioners make—were satisfied with the results: neurofeedback made them feel more relaxed, have more positive feelings, or function better. Moreover, according to users, doing neurofeedback allows them to become, accept, stand up for, rely on, or think as themselves.

This change of the self does not come out of the blue. It is exactly what people have been promised before they decide to do neurofeedback. In newspaper articles neurofeedback is regularly described in terms of self-regulation, self-control, or self-correction. Phrases like 'becoming a completely new human being', 'becoming comfortable in my skin', 'this is really me', or 'I am a better version of myself' are brought up in client reports in magazines or on websites of clinics. Practitioners claim that their clients sometimes talk about 'their new selves' and that parents of their clients speak about the return of their beloved children. An example from a website of a neurofeedback clinic demonstrates this: 'I see my child again, the child from before the depression. That nice, cheerful, social boy who was totally gone! He is back!'[6]

_____

[5] Some users mentioned different problems like tinnitus, anorexia, motor disabilities, or problems due to a stroke. See Appendix 1.

[6] Example translated from www.neurobics.nl/cli-ntenervaringen-depressie/ (accessed on 15-11-2012).

Such statements about 'new' or 'restored selves' and 'people who have come back' are quite common in neurofeedback circles. The author of *A Symphony in the Brain: The evolution of the new brain wave biofeedback*, a book about the rise of neurofeedback in the USA, uses many comparable expressions. 'There's a new person in the house.' 'It's me', says a neurofeedback practitioner to her husband after she underwent the therapy. The same practitioner and her husband had very good results with their son: Neurofeedback 'had given their son—their real son—to them. All this time he had been trapped inside a damaged brain, and now he had been shown a way out.' Later on in the book, another practitioner speaks about a successfully recovered client: 'He was like his old self again, only better' (Robbins, 2000, pp. 104, 103, 129).

This change in the self, so often mentioned on websites and other media, is what people want when they go to a neurofeedback clinic. A neurofeedback user whom I interviewed told me about his search for brain-enhancing devices and therapists, which started in his early teens. He pointed out that this might relate to the fact he had always felt 'different from others': 'I think I had a problem with acceptance and being accepted, and with accepting myself. (…) That's why I thought: maybe I'm not good enough. I have to change. How can I improve, how can I change?' (12). Other users clearly feel they have to change because they do not fit a norm. A woman diagnosed with ADHD, for example, told me: 'I see myself as a very happy girl who sometimes explodes. But these explosions are not convenient in our society. (…) In the future I want to be myself with nice explosions, but without getting into trouble.' She expects that neurofeedback will help her to 'fit the whole' (15).

The wish to improve the self is a clear but insufficient motive for doing neurofeedback—why not, after all, use the much more widely available psychotherapies or pharmaceuticals? Most users who do neurofeedback find this therapy on the Internet or via friends or family because they search for alternative solutions for their problems. Often they are unsatisfied with the general health system, or they do not want to take medication (anymore), because of the side effects. Talking therapies and pills are sometimes also seen as annoying and old-fashioned, while measuring and manipulating the brain is seen as objective and direct. Furthermore, neurofeedback is supposed to be harmless; 'it doesn't hurt to try' is a recurring phrase. One user explains: 'I don't have to believe in it, I just want it to work' (15), and on the Internet someone clarifies: 'I've tried everything, so why not this?' An additional and not unimportant reason to choose neurofeedback is that it seems to offer a solution requiring a minimum amount of effort. To do neurofeedback, people can simply sit and watch a movie or do a racing game, almost as if they are not doing anything. As formulated by a practitioner during an open house of their neurofeedback clinic: 'You don't have to do anything, you can just watch a movie. We are like a home-cinema without pop-corn' (8). However, as will be demonstrated in the next section; doing nothing is not really nothing.

The Process: Enacting the Mind–Body Problem

In his popular science book *Mind Wide Open*, the American journalist Steven Johnson describes his experiences of a neurofeedback session. The practitioner tells him before the sessions starts: 'If I train you too low, you'll feel a little stoned, a little drowsy—you might not want to drive, (…) If I train you too high, you'll be bouncing around the room.' She puts electrodes on the journalist's head, shows him the EEG and says: 'This is you.' And he realizes: 'By changing those thresholds, she can indirectly change my internal states.' When the game begins, he continues: 'I stare at the Pac-Man and wait a few seconds. Nothing happens. I try altering my mental state, but mostly feel as though I'm altering my facial expression to convey a sense of active alertness.' Then his effort is rewarded by a move of the Pac-Man and some beeping and he continues: 'I don't really feel any different but I remember Othmer's [the therapist's] mantra—"be pleased that it's beeping"—and so I try to shut down the part of my brain that's focused on its own activity, and sure enough the beeping starts up again' (S. Johnson, 2004, pp. 101–103).

This book sketches a picture of neurofeedback that separates the person from his or her brain and body. More than that, it implies that the brain is more powerful than the person him- or herself. Someone's performances, which are very conscious and real for the person, can be enhanced by a machine which trains the brain's unconscious. When you do your best, the Pac-Man will not move, but it will when you 'shut down the part of your brain that is focused on its own activity'. In other words, to become good at neurofeedback, you have to submit yourself to your brain. The same conclusion can be drawn after reading advertisements and magazines on the subject. An announcement for an article in a Dutch popular magazine illustrates this: 'Why should you still go in for therapy when you can also send your brain in for treatment? [...] Not you but your brain plays the game, by producing the right brainwaves. The instruction to you (that is, to your consciousness) is just to sit there and not interfere.'[7] Apparently, it is not you, but your brain that does the work.

I asked several practitioners what people have to do during a neurofeedback session. The answers were vague, and sometimes contradictory. One practitioner told me: 'The client should make his brain available, and [he should] consciously follow the learning process of the brain' (7). Another practitioner explained: 'You should let your brain search' (5). A colleague said: 'You need something like relaxed alertness' (6). Sometimes practitioners ask their clients if they 'have control', and in other clinics statements like 'You don't have to do anything' are the norm. So, according to practitioners, people should let their brain search and simultaneously watch its search process. For this they need to be relaxed and alert; they should try to get control, or not do anything at all. All practitioners agree that if one is too aware of the process, it does not work.

This ambiguity of performing an active as well as a passive role is also present in the statements of clients. Some of them describe neurofeedback with phrases like 'You don't have to do anything', or 'It happens all automatically in your head', while other clients explain that they 'have to concentrate, but not too much', 'devote oneself', or 'have to cooperate'. One interviewee clarifies why doing neurofeedback is not as simple as doing nothing: 'It is difficult. Not to concentrate on something you are already aware of, is really hard. If I would ask you not to think about a green apple for the next 30 seconds, this would be a very difficult task' (14).

This struggle to stop thinking what you are thinking, to consciously let something unconscious happen, and to focus on your relaxation appears to be the state for doing neurofeedback. However, one might wonder

---

[7] Announcement for (Mieras, 2004), accessed in June 2009; www.neurocare.nl/nl/node/96

what exactly happens during this fight. Listening to neurofeedback practitioners and clients gives the impression that a struggle between the (conscious) self and the (unconscious) brain takes place. So, although neurofeedback is often described as a conditioning process in which the participant learns to react on the feedback, it appears not to be the subject who conditions him- or herself, but the subject's brain that is conditioned by itself (and not by the self). The role of the user, or the self, is to actively become passive: do not interfere too much, just make your brain available. That is to say, the process is mainly a brain process and occurs mainly at the level of the unconscious. One practitioner offers an interesting metaphor to his clients: 'You travel by bus, but this time you are not the driver like you are used to be, but the passenger. You only have to look out of the window' (7).[8]

Some users, in contrast, make it very clear that they are the driver of the bus, for example, by using their will power to control the process:

> Your EEG has some kind of, well, random fluctuations in the amplitude of the brainwaves. I'm watching these, so I have feedback. Then I use my will power to decrease the amplitudes within that chosen bandwidth. Every time when I see the band goes down I give myself the feeling 'That's what I want, I want that band to go down' and when it goes up I think 'No, that's not what I want.' So, according to me, the will power is central. It is basically a method of self-confirmation. (12)

This quote clearly illustrates the feedback of neurofeedback. The role of this user is active; the will power forces the brain to do the right thing, and the roles of the computer and practitioner are side issues. These are just tools to give the user the feedback from his brain.

However, quotes of another user illustrate that it can also work the other way around: 'The computer trains the brain, or the computer generates the noise, and your brain makes sure that the noise stays away, because I want to listen to the rest of the music. The brain has to work very hard so that I can listen to my music.' In this quote the brain is

---

[8] When explaining his own training process, however, this practitioner attributes himself a more active role: 'I controlled the dominant frequencies in that area. Or, I imagined I would and gave my brain the assignment to manage this and to make sure that I didn't need to pay attention anymore' (7).

trained by the computer because the user wants to finish his music. This user makes a clear distinction between his brain and his self, something which comes more to the fore in the next quote of the same user: 'If I don't pay attention it goes well for a while. It seems that at the moment you start focusing, your brain interrupts with: "Hey, I don't want this signal to be changed." And if you don't pay attention it says "Come on, let me do something again"' (14). In this phrase the user expresses how he argues with his brain during the neurofeedback. When he actively tries to change the computer signal, his brain interrupts. The computer and the brain seem to empower the user, something which becomes even more obvious when the user tries to sabotage the process:

> You are listening to certain sounds, and suddenly your brain starts to stutter and some noise interferes with the music. I tried to sabotage this by thinking about something else and by reacting in a contrary way, but still certain waves decreased. Such a computer can switch over to something else so that what has to be trained will be trained. (14)

That is to say, when the neurofeedback user tries to resist the feedback, the computer trains the brain anyway; against the will of the user.

In the explanations of another user, the distinction between the brain and the self is so clear that the transition is somewhat unexpected. This person starts with 'you', continues with 'your brain', slips into 'it', changes into 'you', and suddenly ends with 'I':

> You are watching traces going up and down. And then they [practitioners] say: 'Well, this is the norm and if you exceed it, we stop the film.' And finally, your brain won't peak out any more because [it thinks]: 'Oh, well, when I do that, the screen is frozen.' So, I—it has to react differently. Deal with another stimulus. (...) So afterwards, you can watch a brain activity scheme with peaks ending up in tranquility. And I am very tired and very hungry. I can eat a whole loaf of bread at such moments. I hate bread, but then I can eat a whole loaf of bread. (15)

Most users I encountered seem to prefer to talk about 'you and your brain' instead of 'me and my brain', but in the former quote the transition into 'I' is very abrupt. This user suddenly switches from 'a' tranquilized brain map into her private psychological state. And she expresses surprise about herself by declaring that she actually 'hates bread'. This user clearly distinguishes herself from her brain, but at the same time she is fully aware of the connection between the two because her brain can change her normal being. In a later part of the interview she wonders: 'What more has changed [in myself] without me knowing? And will I ever regain it?' And she expresses her worries about neurofeedback practitioners: 'You just hope that they have the best intentions' (15).

Doing neurofeedback appears to engender a struggle between the user (whether or not using his or her will power), the brain, the practitioner, and the computer. Who is in charge is a difficult question and varies between the persons asked, but what all cited users have in common is that they bring up a brain besides the self. One user phrases this distinction so naturally it is almost unnoticeable: 'It all happens automatically, and the brain is trained automatically too' (20). The distinction between the brain and the self is not just a matter of vocabulary. The brain becomes a very clear actor for neurofeedback users. It is an entity that can interrupt you, can change you, can harm you, and can cure you.

Mono, Dual, Triad

'States of mind are systematically changed by swallowing pills or receiving injections. Does this not vindicate the union of mind and body, terminating dualism forever?' wonders the philosopher Ian Hacking in his article 'Our Neo-Cartesian Bodies in Parts' (Hacking, 2007, p. 101). He refers to the neuroscientist Antonio R. Damasio, famous for his books *Descartes' Error: Emotion, Reason, and the Human Brain* (1994) and *Looking for Spinoza: Joy, Sorrow, and the Feeling Brain* (2003). *Looking for Spinoza* has been translated in several languages as 'Spinoza was right' and with this, Damasio presents himself as a clear monist. According to Hacking, however, neuroscientists like Damasio are not monists, but 'trialists', since they have created a 'neurologically nested triad' of 'mind, brain and body'. Hacking illustrates this 'three-level scheme' with the example of psychotropic medicines: 'We put pills into our bodies that affect chemicals in our brains, and then we feel better—a state of mind'

(Hacking, 2007). In another article, Hacking writes: 'Within the human organism of flesh and blood, one part, the brain, monitors the body, and another part, the mind (still flesh and blood), monitors the brain and its monitoring of the body' (Hacking, 2005, p. 165).

As demonstrated, such a triad of mind, body, and brain is obviously present in the way users of neurofeedback explain their therapies. By referring to their brain besides their mind and body, they—of course—do not create a different material entity. However, doing neurofeedback makes people so aware of their brain that it becomes a very important and vital entity; the brain starts to perform. People compete with their brain to do neurofeedback and sometimes they have to submit themselves to their brain to succeed in neurofeedback. Yet, this does not erase the self. It is even the opposite; working on the brain (an act of the self) seems to give the self a certain form of autonomy over the brain, as demonstrated by users who state that neurofeedback made them themselves again, or made them finally forgive themselves. So, although neurofeedback can be described as a literal enactment of the mind–body problem, users do not see this enactment as a problem, but as a relief. They do not struggle with the relationship between their selves and their brains, but with the question of how they can make them interact best. Hence, doing neuro-feedback creates a self with a body, a mind, and a brain.[9]

Moreover, the more users are convinced that their problems are located in their brains, the more they seem to confirm their selves as separate entities from these brains. One woman who is very clear about the 'fact' that her problems are brain problems explains how this message did upset her because she realized that this means she is not controlled by herself:

During a course 'ADHD in adults' there was a picture demonstrated with 'these are neurons, this is what they do in normal people and this is what they do in you. (…)' This was beautifully explained with clear images and it made a huge impression on everybody, because it showed that something is wrong. Instead of 'please behave yourself and act normal', this was the

---

[9] Although neurofeedback is a brain training, and performed by the mind, the body is also impor-tant. Users' bodies become literally fixated with electrodes that also register their muscle movement. This can make them very aware of their body. See my own experience in the introduction of this chapter, and my explanations of neurofeedback as a dance of agency in Chap. 6.

evidence that something is wrong in your head. (…) I burst out crying as soon as I arrived home. I thought it was terrible, because I had always thought I was controlled by myself. And that is something completely different than a computer-animated picture: 'Look, this is how it works.' (15)

Another user continuously describes his behavior and problems in brain terms, but hesitates when I ask him if he thinks he is his brain. Instead of splitting himself in brain parts and neurotransmitters, he now splits himself in a 'feeling' part and a 'mental' part:

This is a question…, it even makes me sad…, because for me, it is really, it is such a mystery and I would find it such a pity if it is true what I'm saying. So, there is also one part of me that doesn't want to see it at all in that way. And as I said, I think I live more in a mental part for security reasons, or whatever it is in myself, than in that feeling part. And that feeling part would like to view the world somewhat less rationally and it refuses to see myself as nothing more than a bio-organic robot. (12)

Apparently people want to be their brain to get rid of their problems, but they don't want to be their brain when it reflects on themselves. Or, to phrase this differently: realizing that you are your brain, or are steered by your brain, accentuates the self.[10]

Other Entities Moving Around

Doing neurofeedback requires a split between the self and the brain,[11] but in this process several other entities emerge and start working upon the person's ideas, lives, and feelings. Most users whom I interviewed agree with the practitioners that their problems are brain problems. They have taken over the terminology of their practitioners and explain their problems in phrases like 'my 6 hertz mystery', 'my theta', 'those

---

[10] My own experiences in feeling somewhat endangered in my sense of self, due to my confrontation with my brain, emphasizes this idea.

[11] That is, for those users who are aware of what they are doing. Some people (and probably most children) keep it really simple and state that they are only watching a movie.

alpha and theta things', 'explosions in my brain norm', 'the lack of a certain substance in the brain', 'brain tracks', 'neurons', 'my brain is out of balance'. These brain-related entities are not only explained to them by practitioners (or by books, articles, or teachers), but they are made visible in graphs, diagrams, or other figures that represent their fluctuating brainwaves. In this way brain entities do not only become visible and present in the sense that someone can see them and point to them; they become lively and performative in the sense that they offer people a training goal. One user explains why he keeps on doing neurofeedback: 'I want to see this good state. I want to record it. (...) I want to measure it so that I know "I am rid of my theta", because then I know the crux for my recovery. I have to find the key' (12).

Other entities that can emerge in the neurofeedback process are the colored spots made visible by a qEEG or other brain map. Just like alpha, beta, and theta waves, these yellow and red spots can demonstrate what the problem is (something is wrong in my head), give a training goal (the yellow or red spots should turn green), and can give a feeling of recognition. One client reports on a website: 'My qEEG made clear that some areas could be improved. My anxiety was recognized by a computer: that was something! (...) In the end, the qEEG demonstrated that my critical areas were nicely colored green. On paper, my brain worked much more balanced now.'[12] Sometimes these spots do not only confirm the feelings of the client, but they also give an explanation, as one interviewee explains: 'With me you saw, well, a totally yellow spot and this indicated a depression' (13).

Spots, peaks, waves, and other brain entities that emerged while doing neurofeedback do not stay in the neurofeedback room, but start to intervene in people's personal lives and histories. Users who are done with neurofeedback sometimes claim they are able to control their own brainwave activity, others feel relieved of the control by their brainwaves (I am in control now), or feel in control with their brainwaves due to the feedback (I am balanced). One woman reports that neurofeedback made her less emotional when watching TV and clarifies: 'It looks like my brainwaves automatically take another route, instead of taking the

---

[12] Translated from www.neurobics.nl/cli-ntenervaringen-stress/ (accessed on 9-Mar-2012).

side of the deep emotions' (17). Another client explains that she now understands how her brainwaves were disturbed in her youth: 'In my teenage years, my brainwaves were disturbed by my father who caused a lot of trouble (alcoholism) during the day, evening and night. This made me alert continuously and also during sleep. You can find this in my qEEG. I understand how it works, now' (22). Other users make statements like 'I see the deviations in my brainwaves as the cause of my differently functioning head' (19), or 'My brain received quite a blow [due to a psychosis], and it didn't stop waving in my head'.[13]

Entities that emerged in the neurofeedback process start to infiltrate in people's selves, problems, and daily lives. However, in these new territories, they often encounter other entities. One client reports she is nowadays consciously aware of her brainwave activity, but she also states: 'My thinking is not used to my brainwave activities, yet' (22). When I ask the man who was confronted with the totally yellow spot if he sees his brain as the cause of his problems, he answers: 'Well, I think it is more my life that made me quiet' (13). One practitioner, for whom neurofeedback did not work very well, claims that he is not the kind of person for neurofeedback because his life is too chaotic[14] (6). These kinds of mixtures between psychological and neurological explanations are to an extent present in all interviews. Some practitioners combine neurofeedback with psychotherapies, because they do not want their clients to become dependent on the device, nor on their physiology. Most users speak about their psychology, mind, or psyche besides their brain. One user explains she did everything she could to adapt herself, and the only thing she cannot do by herself is work on her subconscious; this is what she needs neurofeedback for (15). Another user explains his burnout is definitely caused by his brainwaves, but also by his problem with saying 'no' (16), and some users think neurofeedback would never be sufficient to overcome their problems, for example, because 'it is hard to readjust your psychology with this method', or 'what you do and how you live is too important' (14).

---

[13] Translated from www.neurobics.nl/cli-ntenervaringen-stress/ (accessed on 9-3-2012).

[14] Another practitioner who cannot help himself with neurofeedback explains this with 'I am a man; I am not so aware of my body.'

Apart from the interaction between brainwaves and lives, other psychological entities like personality, character, or self-esteem are brought up, as well as various other biological, pharmaceutical, or evolutionary explanations. One user concludes: 'We are primates who can be trained' (19). Another user clarifies his choice for neurofeed-back instead of psychotherapy with 'If you suffer from ADHD you lack a certain substance in your brain' (14), and some people make serious attempts to make one story of all collected explanations for their problems:

> I think my problems are based in my earlier way of coping: working more than 100 hours a week, run half-marathons, not managing my feelings, and so on. As a result, I started to think in the directions of my central nervous system. In my case this was a long-term overburdening of the sympathicus, in such a way that the para-sympathicus stopped functioning properly. I think this also influences hormone regulation, neurotransmitters, and so on. For a layperson it is actually far too complicated, but I think this will influence the brainwave activity. If this activity could have caused this behavior? It could be, I don't know. (17)

Neurofeedback users creatively constitute their own neuro–bio–psycho–social selves. However, they are not the only persons creating such an assemblage. Some researchers have demonstrated, for example, how people diagnosed with ADHD (Bröer & Heerings, 2012) or personality disorders (Pickersgill, 2011), or who are in any other way confronted with their neurological substitution (Martin, 2010; Pickersgill, Cunningham-Burley, & Martin, 2011) use a mixture of neurological and social explanations for their behavior. What makes neurofeedback users specific, however, is the struggle they perform between the self and the brain. The fact that neurofeedback is really an act of doing something—trying to control, balance, or change brain activity by means of concentrating, cooperating, or whatever, instead of undergoing a treatment or diagnoses—makes it a different technology which constitutes a different self. This self is not only an assemblage of neurological, psychological, biological, and social explanations: it contains a clear split between the brain and the self.

Cyborgs and Spirits

Foucault described how people since antiquity had used techniques to improve themselves. Most of these techniques, however, were 'mental' techniques (praying, meditation, confession) and were used to change the mental self. As Pickering phrased it, these technologies of the self produced a 'distinctly human, self-controlled self—the kind of self that sets us apart from animals and things' (2008, p. 5). Neurofeedback users, however, rely on technical devices and want to change their material selves, as, for example, expressed with a shift in brainwaves. As a result, the self they produce is not just a distinctly human self, but a self that is an assemblage of all kinds of material entities, like brainwaves, spots, and peaks.

Moreover, the technical aspects of neurofeedback also have their influence on the users' selves. To explain their problems and the accompanying solutions, neurofeedback users give many computerized and mechanical explanations. They describe the neurofeedback process in terms of 'a defragmentation of your computer', 'cubes put in the right order', 'a computer wiring me', 'a re-programming of my brain', 'my system is unstable', 'my systems resets itself over and over again'. Besides this computer terminology they use other mechanical metaphors to explain their therapies, for example, by saying that neurofeedback tunes, fixes, or wires the brain. Or they make statements that neurofeedback 'puts brakes on the race-car in my head', 'reduced my frequencies by 40%', ensures that 'the right signals reach the right part', puts 'a speedometer in the brain', or trains the brain 'to run at a cruise-control speed'. That is to say, neurofeedback is not only a therapy that produces a materialistic mode of being, but also a technological.

This materialistic and technological form of self-understanding, however, is sometimes combined with a spiritual mode of being. As I have already discussed in Chaps. 2 and 3, and as also became clear in the introduction of this chapter when I explained my own experiences (with Kurzweil and Tolle), neurofeedback has some spiritual connotations. Several clients and practitioners combine neurofeedback with yoga, meditation, or hypnosis. The claim that neurofeedback is like doing meditation or that people who are good at meditation or yoga are also good at neurofeedback is repeatedly phrased. And some of my inter-viewees switched from neurofeedback to meditation techniques, since it appeared to cause the same effect at a much lower cost. It also occurs the other way around; that people start with yoga or meditation but do not experience results fast enough. One practitioner who intends to use neurofeedback on herself phrases it like this:

> I did some work in mindfulness meditation. And recently I started doing yoga. And so, I can influence my brain state to some extent, but the reason I want to do the neurofeedback is because I want to get the state of my brain in its natural state, which is more balanced. So that I don't have to do so much work to pull it into that. Neurofeedback will get it in a better shape without the need of the work I do with mindfulness. It is much more straightforward. (2)

According to some users, yoga, meditation, and neurofeedback appear to have the same effect, which is changing the brain state. The latter tech-nique, however, is easier to employ than the former ones. So for some people, neurofeedback is a technical and time-saving method to achieve a traditional state of being. Moreover, several users use neurofeedback only temporarily—to learn to control their brainwaves—and when they have experienced how to do it, they can further help themselves with yoga or meditation. In other words, they use this 'technical' technology of the self to improve their traditional technologies of the self.

A New Ontology of the Self

Using neurofeedback as a technology of the self modifies people's selves. This changed subjectivity should not be considered simply as a changed perception, or even confusion: it is a very real change for the users and influences the way they live their lives, think their histories, deal with setbacks, and interact with others. In other words, and phrased in the terminology of Foucault, using neurofeedback as a technology of the self constitutes a different mode of being.

This mode of being does not only affect the subject who decides to do neurofeedback, it also influences how he or she interacts with others.

It is characteristic for this therapy in that it involves others easily, and sometimes affects whole families. Quite a few users started to do neurofeedback because it helped their partners, friends, or family members. Sometimes multiple family members are trained in the same clinic, take the training simultaneously,[15] or go on a neurofeedback holiday together.[16] Practitioners often try out their practices on friends and family, and some train their partners, children, or themselves for specific occasions, like having an exam. In their decision to do neurofeedback, most users are mentally or financially supported by their parents or partners.

For some people, neurofeedback really becomes a way of living. Most practitioners once started as clients and became so enthusiastic about neurofeedback that they bought their own equipment and began their own clinics. One (12) bought the equipment himself, discovered a theta peak, and became so obsessed with this peak (or 6 hertz mystery, as he once called it) that he collected multiple brain recordings in various circumstances and sessions to see if his peak would slow down. He tested a range of brain devices—light and sound machines, cranial electrotherapy stimulation, and several forms of neurofeedback—and sent e-mails to therapists and manufacturers to get information on how he could 'get rid of his theta'. In one of his diary notes that he kept with his recordings, he describes how he drives really fast on the highway by using the words 'vroom, vroom, beta', and the first thing you see when you enter his home is a huge poster of his own brain scan.

Not everyone devotes his or her whole life to neurofeedback, but several users who are happy with the results of their training state that neurofeedback will become the solution for their future problems too. One client who was 'very, very, very much helped' with neurofeedback some years ago started to do neurofeedback again, as a 'precaution measure, so that I won't fall back. Just to be sure' (13). Other users changed their techniques of the self from neurofeedback to meditation or yoga, explaining that these techniques change their brainwaves too.

Brainwaves and other brain entities that emerged in the neurofeedback process influence the way people think about their selves and their

---

[15] One practitioner has a room in which four family members (or other relatives) can train together.
[16] See, for example, www.neurofeedbackholiday.com/ (accessed on 15-11-2012).

problems, and it also does this retroactively. One practitioner did a qEEG course and unintentionally detected a beta peak in her brain map. She showed it to me and remarked: 'Everything suddenly makes sense, now that I have seen that bad beta.' When I asked her what she meant, she added: 'Life has been hard for me, you know' and 'I want to become less crazy' (2). This practitioner is now training herself. Several users claim that they finally understand or can accept their problems, and in their explanations about what went wrong, they often include neurofeedback entities.

Neurofeedback obviously results in more neurofeedback. The technique spreads to friends and family, and occasionally it becomes the solution for future problems too. For some people neurofeedback literally becomes a way of living, for others it makes their problems and behaviors more understandable or acceptable, and often neurofeedback helps people to rethink their lives and histories. Using neurofeedback as a technology of the self does not only affect someone's talking and thinking: it can change people's past, present, and future, as well as those of their relatives. That is to say, doing neurofeedback—at least for some people—constitutes a new ontology of oneself.

The Brain We Do

Literature about the self is often concerned with the questions of if and where the self (or mind, or consciousness) is located (e.g. Noë, 2009; Velmans, 2000). Most neuroscientists represent the self as a brain, or as in the brain. Dualists argue that the mind and the brain are separate things. Hacking argues that neuroscientists actually added another entity, and created a threesome: a neurologically nested triad of mind, body, and brain. How can we interpret the experiences of brain device users and all entities they brought up?

The much repeated monistic statement—'the mind is what the brain does'—is broadly accepted in science, but it ends all discussions and leads all human experiences (and the emerged entities) to the realm of language, and with it to unimportance. The dualistic view is untenable from a scientific viewpoint, because it divides people into something subjective (mind, non-material, nurture, culture) and something objective (brain, material, nature), which means 'a fundamental split in the world that runs through human beings, as a result of which they belong only partly to the natural world' (Derksen, 2007, p. 190; see also Barad, 2003; Latour & Crawford, 1993). The neo-Cartesian view of Hacking is not that we are made of different substances; it rather shows something very important: instead of becoming *less* by neuroscience (reduced to our brain), we become *more*. The possibility of regulating specific body or brain parts makes them part of our world (Akrich & Pasveer, 2004; Hacking, 2007; Latour, 2004; Mol & Law, 2004).

In an article on hypoglycemia, the ethnographer and philosopher Annemarie Mol and the sociologist John Law try to understand how patients who live with diabetes regulate their bodies and what this implies for the body:

> We all *have* and *are* a body. But there is a way out of this dichotomous twosome. As part of our daily practices, *we also do (our) bodies*. In practice we enact them. If the body we *have* is the one known by pathologists after our death, while the body we *are* is the one we know ourselves by being self-aware, then what about the body we *do*? (Mol & Law, 2004, p. 45)

According to Mol and Law, bodies of patients with diabetes act, for example, because they can make people sick. Bodies are also enacted in the sense that people try to avoid becoming sick. In these enactments the active body has 'semi-permeable boundaries' and can incorporate some of its surroundings—injecting insulin can become part of oneself—and can also excorporate actions to the world; for instance, when family members can feel that a patient needs sugar. Following Mol and Law, one can state that neuro-feedback users have a brain and are a brain, but also enact a brain. This brain is excorporated, in pieces, waves, spots, and peaks, at a computer screen, and these entities are incorporated as part of the self. The enacted entities, however, do not only procure an interaction with the person doing neuro-feedback, but they also start to enact—or excorporate—some of its actions to the world since they infiltrate in people's lives, histories, and relationships.

Trying to regulate the mind, the brain, or the body enacts these enti-ties as active participants in the world, instead of reducing them to each other. Hence, to follow Hacking, who argues that neuroscientists are actually trialists, is much more interesting and constructive than the monistic or dualistic version. However, the idea that we are 'neurologically nested triads' suggests that all entities put forward by neurofeedback users actually belong to a mind, body, or brain. One might wonder if this extension with one category solves any of the problems of reductionism, like the endless debates about the existence (or localization) of consciousness, responsibility, moral or free will, or that it gives any more insights in a complicated concept like the self. Does it really matter if we are one, two, or three entities? Moreover, adding another category raises the problem of how one decides to which part of the triad the entities belong. Do we need neuroscience to determine if the yellow spot is a phantasm belonging to the mind, or a fact corresponding to a brain state? Should psychiatrists decide if someone's depression is a mind or brain problem? Moreover, if we would agree on a scientific classification system that determines how the entities will be divided, would this be much more inspiring or informing than reducing them to the brain?[17]

In *Reassembling the Social*, the philosopher and anthropologist Bruno Latour asks social scientists: 'Is it not obvious that it makes no empirical sense to refuse to meet the agencies that make people do things? Why not take seriously what members are obstinately saying? Why not follow the direction indicated by their finger when they designate what "makes them act"?' (Latour, 2005, p. 235). His advice is simple: 'Follow the actors themselves or rather that which makes them act, namely the circulating entities' (2005, p. 237). Following Latour would certainly do justice to the experiences of the users. The entities that emerge in their explanation of their self-improving acts are not just in someone's mind, body, or brain; they are out there in the world (projected on a screen), and inside of themselves (steering their behavior). People can interact with them via a computer, take them home as a printed brain map, and let them explain their problems. These entities are performative in the sense that they pop up as language or representations (e.g. a 'yellow spot') but simultaneously

---

[17] According to Hacking, 'neuroscientists like Damasio' have created a threesome. That is, this neurologically nested triad idea is not Hacking's view, but Damasio's view according to Hacking. Hence, my rejection of the triad should not be interpreted as a critique of Hacking, but as an extension of his critique of Damasio's ideas.

become actors (the cause of the depression, the stimulation to do neuro-feedback, the spot that turns from yellow into green).

However, all users in my research had different experiences, used different words, and created different entities. To formulate some conclusion about the subjectivity that is constituted by using these devices, Latour's notion of articulation (2004) can be of help. An articulated entity is an entity that differs from, and is affected by, others. Articulation can be done by scientists, users, graphs, media, manufacturers, practitioners, or whatever makes the entity more visible. The more an entity is articulated, the more alive or embodied (or 'real') it is (Latour, 2004). In the case of neuro-feedback the brain and the self are very well-articulated entities, clearly affecting each other and other entities around. Doing neurofeedback is sometimes called a method of self-confirmation, self-manipulation, self-regulation, self-discipline, and so on, and the goal is obviously a change in the self. To reach this purpose, however, the brain becomes an actor, in the sense that it starts to seek, learn, bring, function, read, know, react, succeed, or understand. In this process—in this excorporation of the brain one could say—all kinds of entities varying from alpha peaks, yellow spots, theta things, 6 Hertz mysteries to bad betas emerge and start having autonomous effects on the person in the sense that they define someone's problems, but also offer a solution. Lives, characters, psyches, feelings, and mental parts, however, are often also embodied in people's selves and histories. Computer metaphors (resetting myself, reprogramming my brain) and spiritual practices (yoga, meditation) are multiply articulated by neurofeedback users, and in media articles other mechanical language (cars, cruise control, speedometer) and sport metaphors (push up for the brain, weightlifting, pump the neurons, train your brain) are much used. That is to say, users of neurofeedback do not only extend their selves with a brain, but they also create an assemblage of neurological, psychological, biological, social, mechanical, spiritual, and other non-classifiable entities. These entities do not always peacefully live together, but often seem to struggle for control. Metaphors of doing sport, resetting your computer, or learning to ride a bicycle stress this striving for control.

New technologies of the self give rise to new selves, Foucault argued, and in the case of neurofeedback this is clearly the case. To do neurofeedback, people excorporate their brains from their selves; articulate them in waves, spots, neurons, parts, and systems; assemble them with all kinds of biological, medical, psychological, computerized, or spiritual entities; and corporate (both in and ex) a private mix of entities to their selves, lives, histories, and relationships. That is to say, neurofeedback users demonstrate that working on the self by working on the brain does not reduce the self to the brain, but multiplies the self with many articulated entities. Or, to quote Latour: 'Reductionism is not a sin for which scientists should make amends, but a dream precisely as unreachable as being alive and having *no* body' (Latour, 2004, p. 226). In spite of all neuroscientific findings, popular beliefs, and commercial aims, the new 'neuroscientific' self can probably best be understood as an extended, assembled, or multiplied self (Brenninkmeijer, 2010).

References

Akrich, M., & Pasveer, B. (2004). Embodiment and disembodiment in childbirth narratives. *Body and Society, 10*(2–3), 63–84.

Amen, D. G. (1998). *Change your brain, change your life: The revolutionary, scientifically proven program for mastering your moods, conquering your anxieties and obsessions, and taming your temper.* New York: Times Books.

Barad, K. (2003). Posthumanist performativity: Toward an understanding of how matter comes to matter. *Signs: Journal of Women in Culture and Society, 28*(3), 801–831.

Bostrom, N. (2005). In defense of posthuman dignity. *Bioethics, 19*(3), 202–214.

Brenninkmeijer, J. (2010). Taking care of one's brain: How manipulating the brain changes people's selves. *History of the Human Sciences, 23*(1), 107–126.

Bröer, C., & Heerings, M. (2012). Neurobiology in public and private discourse: The case of adults with ADHD. *Sociology of Health and Illness, 35*,(1) 49–65. doi:10.1111/j.1467-9566.2012.01477.x.

Churchland, P. S. (2011). *Braintrust: What neuroscience tells us about morality.* Princeton, NJ: Princeton University Press.

Clark, A. (2003). *Natural-Born Cyborgs: Minds, Technologies, and the Future of Human Intelligence.* Oxford: Oxford University Press.

Clark, A. (2008). *Supersizing the mind. Embodiment, action, and cognitive extension.* Oxford: Oxford University Press.

Collins, H. M., Clark, A., & Shrager, J. (2008). Keeping the collectivity in mind? *Phenomenology and the Cognitive Sciences, 7*, 353–374.

Collins, H. M., & Yearley, S. (1992). Epistemological chicken. In A. Pickering (Ed.), *Science as practice and culture* (pp. 301–326). Chicago/London: University of Chicago Press.

Damasio, A. (1994). *Descartes' error: Emotion, reason, and the human brain.* New York: Putnam.

Damasio, A. (2003). *Looking for Spinoza: Joy, sorrow, and the feeling brain.* New York: Harcourt.

Damasio, A. (2012). *Self comes to mind: Constructing the conscious brain.* New York: Random House Incorporated.

Derksen, M. (2007). Cultivating human nature. *New Ideas in Psychology, 25*, 189–206.

Doidge, N. (2007). *The brain that changes itself: Stories of personal triumph.* London (etc): Penguin.

Foucault, M. (1984). On the genealogy of ethics: An overview of work in progress. In P. Rabinow (Ed.), *The Foucault reader* (pp. 340–372). New York: Pantheon Books.

Foucault, M. (1988). Technologies of the self. In L. M. Martin, H. Gutman, & P. H. Hutton (Eds.), *Technologies of the self. A seminar with Michel Foucault* (pp. 16–49). Amherst: The University of Massachusetts Press.

Foucault, M. (1990a). *The history of sexuality: The will to knowledge—Vol. 1.* London/New York: Penguin.

Foucault, M. (1990b). *The history of sexuality, vol. 3: The care of the self.* London/New York: Penguin.

Foucault, M. (1992). *The history of sexuality, vol. 2: The use of pleasure.* London/New York: Penguin.

Foucault, M. (1993). About the beginning of the hermeneutics of the self: Two lectures at Dartmouth. *Political Theory, 21*(2), 198–227.

Foucault, M. (1997a). What is enlightenment? In P. Rabinow (Ed.), *Ethics: Subjectivity & truth, Michel Foucault on truth, beauty, & power 1954–1984* (pp. 303–321). London/New York: Penguin.

Foucault, M. (1997b). On the genealogy of ethics: An overview of work in progress. In P. Rabinow (Ed.), *Ethics: Subjectivity & truth, Michel Foucault on truth, beauty, & power 1954–1984* (pp. 253–280). London/New York: Penguin.

Foucault, M. (1997c). Friendship as a way of life. In P. Rabinow (Ed.), *Ethics: Subjectivity & truth, Michel Foucault on truth, beauty, & power 1954–1984* (pp. 135–140). London/New York: Penguin.

Foucault, M. (1997d). The ethics of the concern for self as a practice of freedom. In P. Rabinow (Ed.), *Ethics: Subjectivity & truth, Michel Foucault on truth, beauty, & power 1954–1984* (pp. 281–301). London/New York: Penguin.

Foucault, M. (1997e). Sexuality and solitude. In P. Rabinow (Ed.), *Ethics: Subjectivity & truth, Michel Foucault on truth, beauty, & power 1954–1984* (pp. 175–184). London/New York: Penguin.

Foucault, M. (1997f). Subjectivity and truth. In P. Rabinow (Ed.), *Ethics: Subjectivity & truth, Michel Foucault on truth, beauty, & power 1954–1984* (pp. 87–92). London/New York: Penguin.

Hacking, I. (1999a). *The social construction of what?* Cambridge, MA: Harvard University Press.

Hacking, I. (1999b). When the trees talk back. *Times Literary Supplement, TLS, 5032*, 13.

Hacking, I. (2005). The Cartesian vision fulfilled: Analogue bodies and digital minds. *Interdisciplinary Science Reviews, 30*(2), 153–166.

Hacking, I. (2006). Making up people. *London Review of Books, 28*(16), 23–26.

Hacking, I. (2007). Our Neo-Cartesian bodies in parts. *Critical Inquiry, 34*, 78–105.

Haraway, D. (1987). A manifesto for Cyborgs: Science, technology, and socialist feminism in the 1980s. *Australian Feminist Studies, 2*(4), 1–42. doi:10.1080/08164649.1987.9961538.

Johnson, S. (2004). *Mind wide open. Your brain and the neuroscience of everyday life*. New York/London: Scribner.

Latour, B. (1993). *We have never been modern*. New York/London: Harvard University Press.

Latour, B. (1999). *Pandora's hope: Essays on the reality of science studies*. Cambridge, MA: Harvard University Press.

Latour, B. (2004). How to talk about the body? The normative dimension of science studies. *Body and Society, 10*(2–3), 205–229.

Latour, B. (2005). *Reassembling the social: An introduction to actor-network-theory*. Oxford/New York: Oxford University Press.

Latour, B., & Crawford, T. H. (1993). An interview with Bruno Latour. *Configurations, 1*(2), 247–268.

LeDoux, J. E. (2003). *Synaptic self: How our brains become who we are*. New York: Penguin.

Lemke, T. (2001). "The birth of bio-politics": Michel Foucault's lecture at the Collège de France on neo-liberal governmentality. *Economy and Society, 30*(2), 190–207. doi:10.1080/03085140120042271.

Martin, E. (2010). Self-making and the brain. *Subjectivity, 3*, 366–381. doi:10.1057/sub.2010.23.

Mieras, M. (2004). Neurofeedback: de nieuwe Prozac. *Psychologie Magazine*. (September 1, 2004): 1410 words. LexisNexis Academic. Web. Date Accessed: 2016/02/17.

Mol, A., & Law, J. (2004). Embodied action, enacted bodies: The example of hypoglycaemia. *Body and Society, 10*(2–3), 43–62.

Noë, A. (2009). *Out of our heads: Why you are not your brain, and other lessons from the biology of consciousness*. New York: Hill and Wang.

Pickering, A. (1995). *The mangle of practice: Time, agency, and science*. Chicago/London: University of Chicago Press.

Pickering, A. (2008). *Brains, selves and spirituality in the history of cybernetics*. Workshop paper, Arizona State University. Retrieved from http://hdl.handle.net/10036/81576

Pickering, A. (2009). Science, contingency and ontology. Retrieved from http://hdl.handle.net/10036/81575

Pickersgill, M. (2011). "Promising" therapies: Neuroscience, clinical practice, and the treatment of psychopathy. *Sociology of Health and Illness, 33*(3), 448–464. doi:10.1111/j.1467-9566.2010.01286.x.

Pickersgill, M., Cunningham-Burley, S., & Martin, P. (2011). Constituting neurologic subjects: Neuroscience, subjectivity and the mundane significance of the brain. *Subjectivity, 4*, 346–365. doi:10.1057/sub.2011.10.

Rabinow, P. (1984). Introduction. In P. Rabinow (Ed.), *The Foucault reader*. New York: Pantheon Books.

Rabinow, P., & Rose, N. (2006). Biopower today. *BioSocieties, 1*(02), 195–217. doi:10.1017/S1745855206040014.

Robbins, J. (2000). *A symphony in the brain: The evolution of the new brain wave biofeedback*. New York: Atlantic Monthly.

Rose, N. (1998). *Inventing our selves. Psychology, power, and personhood*. Cambridge, MA: Cambridge University Press.

Rose, N. (2007). *The politics of life itself: Biomedicine, power, and subjectivity in the twenty-first century*. Princeton, NJ: Princeton University Press.

Swaab, D. F. (2014). *We are our brains: A neurobiography of the brain, from the womb to Alzheimer's*. New York: Spiegel & Grau.

Turkle, S. (2008). *The inner history of devices*. Cambridge, Mass: The MIT Press.

Velmans, M. (2000). *Understanding consciousness*. London: Routledge.

Verbeek, P.-P. (2008). Cyborg intentionality: Rethinking the phenomenology of human-technology relations. *Phenomenology and the Cognitive Sciences, 7*, 387–395.

5

Intermezzo: From Self to Others to Agents

I started to do neurofeedback because I have ADHD. Or ADD, depending on who you meet. And I wanted to get rid of my medication. I don't see myself taking medicines my whole life. [So…] I complained about my medication to my GP and he sent me to several psychological health services. These told me to stop complaining and keep on taking my medication, but they also warned me to quit the medication whenever I would decide to take children. But then what? If I have to quit medication for nine months, I will lose my job, I will start throwing with objects: No one ever thought about that! So, I returned to my GP, being even more frustrated than I was before, and he told me: 'I am not allowed to give you this advice, and I don't know what it does. But here, have a look.' He gave me a bunch of leaflets, and I searched on the Internet to see what it was. I had never heard of neurofeedback, and I found an ADHD-coach working with it. I went there, but he couldn't explain why it would work for me. (…) When I contacted another clinic, they told me: 'Our son did neurofeedback, and he quit his medication years ago.' They could explain very well what happened, and they measured my brain activity before and afterwards, and they were also the cheapest. (15)

This woman gives a clear illustration of the struggle many people experience before they choose neurofeedback. Most users have tried one or

J. Brenninkmeijer, *Neurotechnologies of the Self,*
DOI 10.1057/978-1-137-53386-9_5

more therapies, but became disappointed by the general health care system. This dissatisfaction makes them search for other solutions, which they often find on the Internet or by talking to friends or family members. Their decision to choose an alternative therapy—which they often have to pay for themselves, since neurofeedback is only reimbursed when the practitioner is a registered psychologist (at least in the Netherlands)—can be seen as an act of resistance against the general health care system. It can be seen as a way of tinkering with the possibilities available. The user quoted above wants to get rid of her medication, criticizes the lack of knowledge and explanation of practitioners, and decides to take the therapy that is explained best and costs the least. However, we also see a woman who is diagnosed with ADHD (or ADD) pressed to take medication, sent to many health services, and since she keeps on complaining (resisting), her GP hands her over to alternative non-registered practitioners who persuade her to do neurofeedback with the argument that 'it helped their son too'. These two stories: The individual who takes up the reins and chooses her own path and the system that molds the individual to normal proportions—if needed by handing her over to the alternative circuit—apparently go together perfectly.

Governing Oneself and Others

The ambivalence between the 'free' subject and the disciplining system can also be found in the work of Foucault. Most people know Foucault from his work on power strategies that regulate or normalize the subject—like the prisoner, or the patient. Later on in his career, however, he became interested in those processes in which the person actively modifies oneself (Foucault, 1997d, p. 291). Foucault's later work is sometimes understood as containing a clear split with his former work. Moreover, since Foucault's work on techniques of domination is better known than his work on subjectivity, Foucault's technologies of the self can lead to questions concerning his intentions and consistency. Did he change his mind about the possibilities of individual autonomy and freedom? Did he intend to bring the two perspectives together? How do technologies of the self relate to technologies of domination?

In interviews Foucault explains that after studying 'techniques of production, techniques of signification or communication, and techniques of domination', he became more and more aware that there was another type of technique, which he calls technologies of self (Foucault, 1997e, p. 177). In one of his texts, he also describes how these techniques are related:

> The history of the 'care' and the 'techniques' of the self would thus be a way of doing the history of subjectivity; no longer, however, through the divisions between the mad and the nonmad, the sick and nonsick, delinquents and nondelinquents, nor through the constitution of fields of scientific objectivity giving a place to the living, speaking, laboring subject, but, rather, through the putting in place, and the transformations in our culture, of 'relations with oneself', with their technical armature and their knowledge effects. And in this way one could take up the question of governmentality from a different angle: the government of the self by oneself in its articulation with relations with others (such as one finds in pedagogy, behavior counseling, spiritual direction, the prescription of models for living, and so on). (Foucault, 1997f, p. 88)

According to several scholars who analyzed Foucault's work, this change in perspective from techniques of domination to techniques of self-formation should not be considered as a clear break in Foucault's work. In a collection of Foucault's texts, the editor Paul Rabinow explains that these techniques are analytically distinguishable, but can effectively be combined (Rabinow, 1984). Together with Rose, Rabinow argues that the combination of these techniques—also called biopower—in liberal societies has taken the form of governing life itself (Rabinow & Rose, 2006; Rose, 2007).[1]

Seen from a biopower perspective, neurofeedback is a technology that individuals use to improve themselves under the regime of certain authorities and ways of knowledge. Before people end up doing neurofeedback, their functioning can, for example, be judged as inadequate according to the standards of their employees; their behavior might be diagnosed as abnormal, or disordered, by authorities like psychologists; and their

---

[1] Rabinow and Rose distinguish three strategies of biopower in the work of Foucault: truth discourses, strategies of intervention, and modes of subjectification (Rabinow & Rose, 2006).

brains are probably seen as unbalanced by their neurofeedback practitioners. Doing neurofeedback, then, becomes a way to achieve the norms as
defined in a liberal society. It can be seen as one of many available strategies that make people more aware of—and with this responsible for—
their own health and happiness, which belongs to a (liberal) principle in
which healthy brains are, for example, seen as the crux to generate happy,
hard-working, and reliable citizens.

In this book, technologies of the self are understood as technologies to
care for the self that are, or can be, part of biopower in (neoliberal) societies. One important text for this interpretation is *About the Beginning of
the Hermeneutics of the Self*, in which Foucault declares that to understand
the genealogy of the subject we have to take into account the interaction
between techniques of domination and techniques of the self (Foucault,
1993; see also Lemke, 2001). It is obvious that neurofeedback users do
not only work on themselves by themselves; their wish to improve is
clearly related or steered by others, and by doing neurofeedback they rely
on the knowledge (and demonstration) of neurofeedback practitioners.
To understand the relation between neurofeedback as a technology of the
self and a technology of domination, I analyze the neurofeedback process
in more detail in Chap. 6, where I try to retrieve the important actors that
help the client to constitute his or her new way of being.

From Others to Agents: When the Brains Talk Back

To study what is actually going on in the neurofeedback room, I relied
on interviews and other reports of clients and practitioners, observed
neurofeedback demonstrations, attended meetings for practitioners,
and observed a neurofeedback experiment on school children. During
these interviews and observations, however, several different and sometimes unexpected actors emerged: EEGs acted unpredictably, parents
intervened in the treatment of their children, and computers sometimes
appeared to have the knowledge practitioners lack. In other words,
although neurofeedback is a technology of the self, clearly modifying the
subjects' understanding and experience of themselves, there are many
other actors involved, both human and not human. Brain maps, test

results, and computer programs appear to be just as important as practitioners, scientists, and clients.

Neurofeedback distinguishes itself from other therapies by the obvious role it gives to technological tools, and hence it would make sense to include these 'non-human entities' in my account of neurofeedback. Since Foucault's ideas are focused on human beings and human (power) relationships, I will not just rely on his work to study the role of others. It was mainly after his death that some scholars started to study the impact 'things' or machines have in Western societies, and hence rejected the ontological distinction between subjects and objects. The feminist and philosopher Donna Haraway, for example, argued that our society is full of 'couplings' between organisms and machines, which she called 'cyborgs' (Haraway, 1987). Latour also reasoned that humans and things are so interdependent that they are actually 'quasi-subjects' and 'quasi-objects', and to get rid of the dichotomy he called them 'hybrids', or 'actants' (Latour, 1993, 2005).

In one of his books, Latour gives an example which might clarify his argument. He introduces a man who picks up a handgun (1999). For Latour, the man with a gun becomes a different subject (e.g. a criminal) and the gun in the hand of the man becomes a different object (e.g. a weapon). Since the 'man with the gun' is not only subject, and the 'gun in the hand of a man' is not only object, they are both something else, which Latour calls 'actants'. That is, for Latour, 'It is neither people nor guns that kill. Responsibility for action must be shared among the various actants' (1999, p. 180). Such arguments inspired many scholars who, for example, analyzed how humans merged with technologies such as hearing aids and pacemakers to help them function normally, social software to keep contact with their friends, cell phones to remember their appointments, or brain stimulators that make them feel less anxious or more happy (Clark, 2003, 2008; Verbeek, 2008; Turkle, 2008).

The amalgamation of humans with technologies is often described as a (coming) 'post-human' or 'transhuman' state since human beings can, or will in the near future, not be considered as strictly human anymore but are transforming into man–machine entities, or cyborgs (Verbeek, 2008; Bostrom, 2005). However, there is also criticism of this post-human tradition. Some scholars argue that using tools or technologies is actually a very human thing to do, which makes the addition of 'post' or 'trans' unnecessary. This 'natural' extension of the self is, for example, defended by the philosopher Andy Clark: 'Such extensions should not be thought of as rendering us in any way post-human; not because they are not deeply transformative, but because we humans are naturally designed to be the subjects of just such repeated transformations!' (Clark, 2003, p. 142). Latour also argues that the connection between subjects and objects (hybrids) is nothing new or modern. In fact, he argues the opposite; while in pre-modern cultures subjects and objects are not distinguished and hybrids (for instance, holy trees) are widely accepted, the attempt of modern cultures to distinguish subjects from objects brings hybrids into being even more quickly (Latour, 1993).

Other authors, however, criticize the idea that non-humans act, or that humans and non-humans amalgamate. Latour's explanation of the man with the gun is, for example, criticized by Hacking in an article titled 'When the trees talk back' (1999b). This title is meant to be ironical. It is obvious that trees don't talk, and Hacking uses this to illustrate that objects have no agency. He writes: 'I am not made a new agent when I simply pick up a gun. The gun is not an agent. There is no hybrid man–gun.' Comparable arguments are made by other scholars. Clark's concept of an 'extended mind', in which cell phones and other objects that support cognitive functioning are seen as part of people's mind, is discussed by the sociologist of science Harry Collins who, for example, wonders if Clark also thinks that he is as much part of his cat's extended mind, as his cat is part of his mind (Collins, Clark, & Shrager, 2008; see also Collins & Yearley, 1992).[2]

This discussion about agency becomes increasingly complicated when we add a brain to the subject. If a human with a gun and a brain injury kills someone, then who is the agent? Imagine that we put the person in a scanner which visualizes the affected brain parts. Is the brain on the screen an agent? A human? A mind or a body; immaterial or material; a subject or an object? Moreover, could Hacking have written an article

---

[2] This is a simplified explanation of Collins' critique. Literally, he writes: 'Clark readily talks of "dovetailing"—extending our abilities in virtual networks, prostheses, and so forth, by fully engaging with them but he does not talk about dovetailing with Lolo [the cat]. A dovetail joint is a symmetrical joint: If you were to dovetail with Lolo, then Lolo would have to dovetail with you but Lolo can't—Lolo's brain does not have human-like flexibility.'

titled 'When the brains talk back', in which he made the argument: 'I am not a different man when I have a damaged brain. The brain is not an agent. There is no hybrid man–brain'? That is to say, although the plea for symmetry between humans and non-humans is often criticized, subject–object boundaries are becoming increasingly arbitrary.

This also becomes manifest when listening to neuroscientists. As discussed in Chap. 3, book titles like *Synaptic Self* (LeDoux, 2003), *Self Comes to Mind* (Damasio, 2012), or *You Are Your Brain* (Swaab, 2014) seem to dissolve this subject–object boundary. However, as this book argues, involving the brain in our understandings of the self does not reduce the self to the brain, but extends the self with a brain and multiple other entities. That is, although neuroscientists seem to work in the opposite direction of other scientists, since they try to get rid of the subject–object distinction that other scientists so obstinately try to maintain, they actually do the same: they create 'hybrids' of brains and selves, and all kinds of other non-classifiable entities (e.g. yellow spot).

Not only does the arbitrariness of the distinction between subjects and objects become obvious in the neurofeedback process, but so does the agency of these interactions. Non-humans like computers and brain maps are crucial for this therapy and also have their impact on the way people constitute themselves ('my system resets itself', 'my anxiety was recognized by a computer', 'that bad beta'), so it is inevitable to include non-humans in my description of the neurofeedback process. However, since my analysis focuses on subjectivity—a human subjectivity, seen from a human perspective—it is also problematic to analyze neurofeedback with a radical symmetrical approach. (Critics could ask me if neurofeedback users are as much part of their brainwaves as their brainwaves are part of them.) Hence, to bring non-human actors like the computer into my account of neurofeedback, I decided to follow Pickering, who combines insights of Foucault, work on performativity, and perspectives of scholars like Latour and Haraway, and describes scientific practices as dances of agency between human and non-human actors. For Pickering the relation between humans and non-humans is symmetrical in the sense, and during the time, that they together perform, but asymmetrical in the sense that humans have intentions (goals, plans) while non-humans do not, and in the sense that we describe practices from a human point of view (Pickering, 1995). He describes scientific practices as a 'constitutive back and forth between human agents who contrive specific material set-ups, and the agency of those set-ups themselves—what they do' (Pickering, 2009, p. 4). That is, the 'dance of agency' seeks to include the performances of humans and non-humans in any given account, without making them equivalent.

6

Neurofeedback as a Dance of Agency

Participant (11): How about homunculus? What is it that is doing that?

Course supervisor (4): We think it is the connections (…) Silent synapses (…) Getting the network in the right state.

Participant (11): Is it the mind? What is it that makes A into B? What is the you?

Course supervisor (3): [Laughs] Well, let's start with Descartes… (…) The brain *produces*… (…) The brain *knows*… (…) I don't see the problem. When you work too hard and you are very anxious, and when you stop working so hard you can become less anxious. The mind *is* the body.

Participant (29): I see it like a plastic band; if you stress it too often it stays this way. It is like training a muscle.

Course supervisor (3): Some people never had the chance to experience these brainwaves. So you are training them to produce this brainwave.

Course supervisor (4): It is like a field of weeds. If you walk through it once, and a few weeks later you do it again, you would take a different route. But if you do it very shortly after the first walk, you would take the same way. And like this you can create a path.

Participant (11): This is a very useful metaphor. I think I can use it for the parents of the children I treat.

Course supervisor (3): He is very good at metaphors; he has a good parietal lobe.

© The Editor(s) (if applicable) and The Author(s) 2016

J. Brenninkmeijer, *Neurotechnologies of the Self*,

DOI 10.1057/978-1-137-53386-9_6

This conversation, noted down during a neurofeedback course for novice practitioners where I was allowed to make some observations, is a discussion about agency. Participant 11 wonders what it actually is that responds to the feedback. The answers he receives are materialistic and mechanical; it is not you who is doing the neurofeedback, but the connections, synapses, network, brain, plastic band, a field of weeds. Moreover, it is not even the course supervisor who is good at metaphors, but his parietal lobe. In other words, the agent of the process—the entity that performs—is material and multiple.

A comparable phenomenon was illustrated in Chap. 4. Neurofeedback users came up with all kinds of entities that struggled and collaborated in the neurofeedback process. Selves and brains (presumed to coincide) were split and had to interact; neurotransmitters, spots, and brainwaves emerged and became the cause of and solution for the problems; and social circumstances and personal characteristics were produced and controlled by these spots and waves. In this process, which is often expressed as a method for attaining self-control, it is difficult to explain who or what has control. Practitioners ask their clients if they have control. Most clients, however, express that they do not feel control over but feel controlled by their brain, the computer, or the practitioner. Some clients explain that neurofeedback teaches them to attain (some) control over their brainwave activities, but others express a feeling of losing control, since they feel steered by their neurons, brainwaves, or other brain entities. That is to say, doing neurofeedback constitutes a self that exists of many entities that struggle around for control. To find out how all these entities emerged it makes sense to describe the neurofeedback process in more detail. This chapter applies Andrew Pickering's ideas and analyzes neurofeedback as a dance of agency between human and non-human actors who struggle, collaborate, and swap roles in a process which creates a new self for the neurofeedback client.

Dance of Agency

One of the central characteristics of Pickering's dance of agency is that it entails a process of 'tinkering' (Knorr-Cetina, 1981), 'bricolage' (Latour & Woolgar, 1979) or, in Pickering's terms, 'tuning'. In his book *The Mangle of Practice* (1995), Pickering, for example, demonstrates how the quark was constructed through a process of trial and error, by presenting a scientist who assembles a setup, stands back to see what happens, reconfigures the apparatus, sees what happens, reassembles the setup, and so on:

> As a classic human agent, Morpurgo assembled his apparatus, switched it on, and then, surrendering his active role, stood back to watch what would happen—literally, through a microscope. Swapping roles, the material world was in turn free to perform as it would: the grains levitated and moved away from their equilibrium positions when the electric field was applied. And immediately a problem arose. The very first grain acted strangely. (Pickering, 1995, pp. 79, 80)

Important in Pickering's ideas is that this acting 'strangely' of the grain should be taken seriously and not symbolically or semiotically. Pickering proposes a shift from an epistemological to an ontological way of thinking (Pickering, 1995, 2007, 2009, 2010). He criticizes modern scientists for their representational approach in which they leave no space and even veil the performative aspects of our world.[1] Instead, Pickering proposes a performative idiom for thinking about science, which describes the interplay between epistemology and ontology. He describes scientific practices as engendering ontological changes and gives examples of humans and machines together performing 'ontological theater': they play (or dance) together and bring new forms of being into the world. Pickering writes that these dances of agency 'conjure up an image of the material world not as fixed, static and knowable, but as endlessly lively. The world performs—does things—[...it] is a place of endlessly emergent performativity' (Pickering, 2009, pp. 4, 5).

Material agents, like the grain, do not only perform, their performances often count as a resistance for the human agent. That is to say, material or other agents (a concept, for example) do not always do what was intended by the human agent. Such resistances induce a new action in the form

---

[1] With the term 'performative' Pickering usually means something like 'having agency', but he also uses it to refer to the capacity of a 'dance of agency' to bring something into being (Pickering, 1995, 2010). See also Barad (2003) and Callon (2007) for discussions about performativity and the shift from a representational to a performative idiom in the natural and social sciences.

of an accommodation: the scientist has to make a revision to his or her strategy, for example, by tuning the setup or changing his or her concepts of the world. In Pickering's words, dances of agency are structured as 'a dialectic of resistance and accommodation' (Pickering, 1995, p. 22).

In addition to these aspects—tuning, ontological theater, and resistance—the metaphor of a dance is also useful. A dance is lively and fluid.[2] Actors are noticeably moving together, but it is not always possible to define who or what is leading at what moment, let alone what will happen next. A dance allows unexpected movements, and as long as there are enough actors, it allows stepping in or out. These characteristics of a dance are also relevant for describing the neurofeedback process since the 'choreography' of this practice is not always very clear, while its liveliness is.

Searching for Feedback

While doing neurofeedback, adult clients mostly listen to music, watch a movie, or simply stare at graphs that represent their fluctuating brain-waves. Whenever their brain produces the right frequencies, they hear a beep and the music gets louder, the screen enlarges, or the bar of the graph goes up. However, about half of the neurofeedback clients are children, and about half of the children are boys diagnosed with ADHD. Children can watch movies or listen to music, but if they want some action, they can also play games in which they, for example, have to speed up a car. In other neurofeedback games children have to make a smiley smile, or let a bear grumble, with their brainwaves.

In this way users receive feedback of their brain activity they are normally not aware of, and can try to influence this. What someone exactly has to do to change his or her brainwaves, however, remains unclear. Neurofeedback clients and practitioners refer to different agencies to explain the process. Clients, for example, appoint their subconscious, their will power, the practitioner, or the computer as the actor that trains their brain. The statement 'you don't have to do anything' also recurs in client interviews. Practitioners, on the other hand, use a variety of meta-phors to describe the process: someone refers to buses, where the client is

---

[2] While a network, for example, appears more fixed and static (See also Kendall & Michael, 2001).

the passenger and the client's brain is the driver; others describe it as akin to learning to ride a bicycle; one practitioner evokes childhood 'warmer and colder' games; and in the introduction of this chapter a course supervisor describes it as creating a path in a field of weeds. Expressions like 'you don't have to do anything, it is the brain that does the work' are repeatedly used by practitioners as well.

During a neurofeedback course for novice practitioners, several games were demonstrated. One of these was a caterpillar game in which the client has to speed up three caterpillars representing his or her theta, beta, and SMR (12–15 Hertz) frequencies. Playing this game is actually playing a competition between your own brainwaves. The client is connected to three electrodes—one on the scalp that measures the brain's activity and two on the earlobes to ground these measurements—and watches a screen with a pink, a green, and a blue caterpillar moving forward. The practitioner, at the same time, watches a computer screen and sees fluctuating brainwaves. Based upon these fluctuations (and the chosen protocol), the practitioner may decide, for example, to give feedback when high beta decreases and theta and SMR increase. She (or he) tunes the feedback thresholds and can give the assignment: 'Watch your blue caterpillar; they all have to speed up.' Thereupon, a period of passively waiting begins for her. The client, sitting at the other side of the computer, sees (perhaps) that his (or her) blue caterpillar has decelerated, and has to do 'something' to make it speed up again. He tries to concentrate, relax, focus on one point, or do nothing at all, and waits to see if and when the blue caterpillar will accelerate. When it does, or if it takes too long, the practitioner can become active again. She can, for example, decide that beta is reducing, but that theta and SMR are still not at high enough levels, and she changes the threshold bars again, and sits back to see what will happen now. The client, who probably thought he had mastered the neurofeedback training, suddenly notices that the pink and green caterpillar, representing theta and SMR, stop moving, and has to do something to make them speed up again.[3]

---

[3] This example is based on the practices as demonstrated during the neurofeedback course for novice practitioners. More common, however, is to decide how (e.g. SMR up, and theta down) and where (e.g. on c4, right central) to train before the session starts. Other games, like Pac-Man, or a racing game, are also used more often than the caterpillar game, and for adults watching a movie with a fluctuating screen is the most common. However, the caterpillar game is interesting because it illustrates a competition between someone's brainwaves.

In turn, and from different angles, the client and the practitioner initiate a dance of agency that takes the form of 'a dialectic of resistance and accommodation' (Pickering, 1995) in which the caterpillars (or smileys, or sounds), the brainwaves, and the human actors are alternately passive or active. This tinkering to make the feedback work, however, is only one part of the story. To make the therapy as a whole a success, many more actors are involved that struggle, collaborate, and swap roles. In the following I will trace the other, less obvious actors that are important in the neurofeedback process, from their help in creating and motivating the client, to their performances during the training and their assistance in collecting the results.

Creating the Client

Although neurofeedback can be something people do 'on their own'—for example, with their own neurofeedback devices at home—the help of 'others' is usually very important, such as that offered by practitioners who promote this form of brain training, or parents who characterize their child as abnormal. As Foucault demonstrated, this often well-intended 'help of others' also regulates the behavior of the individual. That is to say, neurofeedback clients do not only work on themselves, they are also 'made up' into clients, and 'disciplined' to improve their brains, for example, by worrying parents, brain awareness campaigns, educative projects, and individual brain maps (Foucault, 2004; Hacking, 2006, 2007; Rabinow & Rose, 2006).

The influence of others becomes obvious when listening to and observing the work of practitioners. Especially in the treatment of children and young adults, practitioners often mention the role of parents, teachers, or psychiatrists who encourage the person to improve. Not infrequently, this encouragement to change coincides with some intimidation. One practitioner, talking about a teenager who successfully recovered from a depression, clearly illustrates the disciplining power of others (Foucault, 2004):

> This client had much resistance against doing anything at all. And neurofeedback just turned the switch. He said this already happened after 3 sessions. (…) The idea was a compulsory admission [in a psychiatric hospital], but the

waiting list was very long. They wanted compulsory admission to stuff him with medicines so to say, just to see if this would… to stimulate him to get up early, to see if it would have any effect. (5)

This client clearly had few options, but also when there seems to be no pressure at all, the wish or urge to improve oneself is generally inspired by others. Several neurofeedback clients, for example, state they would like to change because they want to 'fit in', or because they feel 'different' to others.

However, whilst the influence of others might prompt a desire to change, it does not automatically lead to the choice of neurofeedback as treatment. Before people become neurofeedback clients, they first have to believe it is possible or necessary to work on themselves by changing their brainwaves. They must be convinced that their problems or failures are abnormal—often evidenced by a psychiatric label such as ADHD, depression, or autism—and that the cause and solution of their complaints lies in their brain (Dumit, 2003; Rose, 2007). This is not always obvious for everyone, as a researcher examining the efficacy of neurofeedback illustrates: 'People first have to recognize that they have ADHD. They must understand that it is a problem before they want to make such an effort to resolve the problem' (1).

Defining the problem must be followed by defining the problem as a brain problem. Trying to make people aware of their brains is becoming a key part of scientific and popular scientific discourse, and neurofeedback practitioners contribute where they can. Practitioners are often very active on the Internet, they are willing to be in the media, they welcome potential clients into their clinics during 'open houses', and they are happy to demonstrate their practices during educational events such as museum exhibitions. One practitioner offered her help to a science museum where she could demonstrate neurofeedback to the audience. She explains: 'What did it look like? Well, actually I just hooked them up. People love to see their brainwaves. Yes. So it is great if you can teach: "this is a slow wave", "this is a fast one", "this is what we do"' (2).

Making people aware of their brains can also be a one-to-one process. During a neurofeedback course for novice practitioners the participants openly discuss each other's brainwaves. One practitioner remarks about the person who is 'hooked up': 'What nice brainwaves! I like these. Sometimes I really don't like them' (10). When one of the participants explains that he reacts inconsistently to coffee, one of the supervisors responds: 'This means that you have an unstable [brain] arousal' (4). What impact these kinds of 'disclosures' of people's brain activity can have is illustrated by a practitioner who reveals in an interview how she was confronted with her own brain map during an electroencephalography (EEG) course: 'Everything suddenly makes sense, now that I have seen that bad beta' (2).[4]

After making people aware of their brain problems, practitioners have to convince their potential clients that neurofeedback is an effective therapy. For this they often use stories of clients who have successfully recovered, as well as metaphors that symbolize the working of neurofeedback. One frequently used metaphor is learning to ride a bicycle.[5] This metaphor also emerges during an open house organized by a neurofeedback clinic when a woman and her adult son receive information that is too technical for them to fully understand. The practitioner explains the neurofeedback training by comparing it with the training wheels used to teach children to ride a bicycle. The metaphor brightens the faces of the woman and her son, they reply 'oh, yes, so it works like that' and decide to go for an intake session. Another metaphor that is used during the open house concerns signposts in your brain that have fallen down, so that you get lost. Neurofeedback is about putting the signposts back. At first glance, signposts and bicycles have nothing to do with the brain, or with the working mechanisms within it, but apparently this does not make them less effective in convincing clients. The opposite is possibly true: using a metaphor from everyday life seems to make people feel more comfortable with the otherwise incomprehensible therapy and their own mysterious brain.

When scientists or practitioners inform people about their brain problems and potentials, they suggest that people are responsible for their

---

[4] See also Chap. 4.
[5] Riding a bicycle is a much used example in the discussion on tacit knowledge; knowledge that we have or can obtain, but only in the sense of doing it, not in explaining it. We know how to swim, or ride a bicycle, but cannot explain how. See Polanyi (1962) and Collins (2000).

own brain health.[6] In some cases, practitioners literally expound this message. One of the neurofeedback course participants, a neuropsychologist, explains that he is interested in neurofeedback because it provides a solution, instead of only an identification of the problem. He describes colleagues who simply state that 'broken brains cannot be fixed', while the great benefit of neurofeedback is that it 'gives control back to the client' (11). A comparable message is proposed by a practitioner responding to a question about the future of neurofeedback:

> We got in some kind of mind set in which we handed over our responsibility to the experts. We don't take responsibility ourselves; we go to a doctor and take medication rather than change ourselves. And I have a feeling that this might be going to change. I think there is something in the air, there is a shift going on. That people want to take responsibility for themselves. (2)

In other words, neurofeedback practitioners emphasize the message that people are responsible for their own brains and happiness. According to them, stabilizing your broken brains with a pill is not taking enough responsibility. People should change themselves by their brains, by themselves.

In the creation of a neurofeedback client many actors are—intentionally or unintentionally—involved. Parents, psychiatrists, and diagnoses help the client to become aware of his or her problems. Museum demonstrations, campaigns, and brain maps make the client aware of his or her brain. And practitioners, metaphors, and success cases make him or her aware of the solution. In this awareness process the responsible agent 'dances' around; first it is the person him- or herself who is responsible (I am so hyperactive); next the responsibility is distracted from the self and connected to a diagnosis (I cannot help it, it is my ADHD); thereupon, the behavior becomes the responsibility of the brain (it is not my fault, it is my brain); and then the person becomes responsible for his or her own behavior again (I have to take care of my brain).

---

[6] See also Dehue (2008); Rose (2007); Roy (2008) for more examples and explanations of how people become more and more responsible for their own health and happiness.

Motivating the Mind, Body, Brain

As shown above, a whole process precedes the involvement of a client. However, this does not always mean that a client is also a collaborating actor. In some cases, clients, especially children and teenagers, can be hindering and sabotaging actors. One of the neurofeedback course supervisors talks about a difficult client who admitted after many sessions: 'I don't want to be a swot.' According to this practitioner this was the reason why the neurofeedback did not work for the boy: 'He was sabotaging the training. Every session was a fight' (4).

Because of the problem of unmotivated clients, various tricks and tips are given during the neurofeedback course for practitioners to keep clients alert and motivated. Several encouragements are included in the neurofeedback tools: clients can watch movies, listen to their favorite music, or play amusing games. For children, there are special toys. During the neurofeedback course the trainers demonstrate the neurofeedback 'Jedi' helmet which makes it possible to lift a plastic ball with your brainwaves, and a teddy bear, called 'Neury the Bear', which gives a rewarding sound—snoring or growling—when the connected child produces the right brainwaves. Other tips the supervisors provide include the use of watches that buzz every three minutes to keep children attentive and for strongly unmotivated children one of the supervisors advises: 'Put a PlayStation on the neurofeedback device and say "If you have so many points, this PlayStation is yours"' (3).

These kinds of tips and tricks are helpful and sometimes necessary to keep clients motivated, but practitioners use many more strategies to instruct, relax, and stimulate their clients. When they talked to each other or to me, the language that they used was very mechanical: connecting the client to the EEG and the computer was often described as 'hooking up', tuning the frequencies of the computer program to change someone's brain activity was expressed as 'screwing up or down', clients were sometimes described as people with 'broken brains', and successfully treating a client was commonly expressed as 'fixing' someone. However, these words are usually not employed when the client is around and one practitioner even asked me to delete the word 'fixing' from his interview text.

Practitioners do not tell their clients about their broken brains that will be fixed by hooking them up to the computer and screwing the frequencies in their head up and down. The opposite is true: practitioners calm down their clients, talk gently to them, and make use of several techniques to keep the person motivated. This is because, as one practitioner phrases: 'You are not only working with the brain, you are working with the whole person' (2). That practitioners indeed work with 'the whole person' is clearly demonstrated in interviews with the researchers of, and observations during, a neurofeedback experiment with eleven-year-old children. The researcher explains how he keeps his subjects at ease ('I just tell them that they are going to listen to some beautiful sounds') and what they do when their subjects are not:

> [One child] is afraid of doing neurofeedback. [The experimenter] has to walk with him around the building, let him talk to children who already did neurofeedback, and really comfort him. He is a very nervous kid who worries a lot, and he didn't produce any theta waves yet. (1)

Other children participating in the experiment are also comforted, encouraged, and corrected. When one child worries about her brainwave pattern: 'Is it good? Is it bad? Is it flat?' the experimenter responds: 'No it is fine. Do as you said: find the happy feeling, like when you are singing' (2). During and in between the neurofeedback games, the experimenter continuously intervenes with utterances like 'Woah, that is a lovely one!', 'A great start!', 'That looks absolutely fine!', 'You are really in the zone, aren't you?'. The experimenter comforts the children and keeps their attention in the right place, but in between she also corrects their posture: 'Could you put your feet flat on the ground?', 'You must sit still', 'Keep relaxing', 'Nice sitting', and during a different session she tells the children to 'Lay down on your back' (2). Apparently, practitioners do not only work with the brain, and the mind (or the person), but also with the body. To phrase this differently, to train someone's brain, practitioners first need to calm down the mind and correct the body.

Instructing clients, especially children, seems to require more than just a simple explanation. Besides being instructed in how to perform neurofeedback, participants also have to be stimulated, corrected, and reassured. Many tools are developed to keep the attention of adults and children, and psychological strategies can also help to motivate the client. These techniques demonstrate that neurofeedback is a matter of not only the head or the brain, but also the mind and the body: clients have to be calmed down before the training and motivated during the process. Furthermore, they have to sit up with their feet flat on the ground, or lay quietly back in darkness with their feet on a footstool.[7]

To phrase this in Pickering's terms, before the brainwaves of neurofeedback clients do what the practitioner wants, there is a lot of *resistance* from the client's mind and body. To accommodate the mind and the body practitioners need several tools, tricks, and words. So, the 'failure to achieve an intended capture of agency in practice' (a cooperative mind and body) and the 'active human strategy of response to resistance' (using tricks and tools) take the form of 'a dialectic of resistance and accommodation', that is, a 'dance of agency' (Pickering, 1995, p. 22).

Choreography of the Dance

Creating a cooperative neurofeedback client requires many actors, from practitioners and parents, advertisements and diagnosis, to the clients' minds and bodies. Doing neurofeedback is a process of tinkering in which the client actively does something (or nothing) and passively waits to find out what will happen next, and in which the practitioner, on the other hand, also actively tunes the adjustments on his computer and passively waits to see what will happen. In this section I will elaborate upon the process of neurofeedback by focusing on the choreography of the dance.

In the first instance, neurofeedback appears to be a practice between an expert and a subject, that is, a practitioner and a client. However, when interviewing practitioners it becomes clear that they are not always the ones who are in charge of the process. One of them, for example, describes how he started his neurofeedback career using the same protocol for everyone in the same way and confesses: 'The unsatisfactory thing was that I actually had no clue what I did' (7). Nowadays, this practitioner explains, more research is conducted and more information has become available, but still several processes are unclear, such as why neurofeedback only works in some people and not in others, and what the training actually

---

[7] The instructions to sit up or lay back depend on the specific training. For alpha and theta sessions clients should close their eyes and relax; for beta sessions clients should mostly sit up and pay attention.

does on a neuronal level. In spite of these 'puzzling material performances' (Pickering, 1995, p. 82) the practitioner has to depart from 'something' to decide what to train and where to train. This 'something' can vary from standard protocols to EEG results, client's diagnosis, to computer information. None of these agents, however, appears to be very reliable.

Standardized protocols, for example, do not always work. One strategy is to define someone's arousal state by questioning the client about his or her habits. Drinking a lot of coffee, for instance, is seen as a characteristic for underaroused types and drinking too much alcohol for overaroused types. However, one practitioner explains about an experiment in which he is involved:

> [The researcher] prefers people to be an A, B, or C, so that you can apply an A, B, or C protocol. If you are an A you are over-aroused, B is under-aroused, and C is unstable. (…) But it has already turned out that it is not always the case that someone is only over-aroused—he can also have some under-aroused characteristics. (7)

Departing from the client's diagnosis or complaints, on the other hand, does not always have the desired effect either. One practitioner reports:

> You start with what people say and you have this recognition and experience. It is empirical, that you think when you hear some complaints: 'Oh, yes, SMR', or 'Now, I will train beta, or alpha for people with anxiety or stress. So, you need to start somewhere, but it also happens that something doesn't work out. (6)

The fact that protocols do not straightforwardly define what to train, how much to train, and where to train, however, does not appear to be a problem for practitioners. One frequently cited study analyzing the results of many neurofeedback experiments on ADHD concludes that the effects are positive, irrespective of the specific training.[8] When these results are demonstrated during a meeting for practitioners, one of the participants jokes: 'Yes, sometimes I also think: just kicking the brain a bit will also work.'

---

[8] 'Interesting, post-hoc analyses did not reveal any differences between the different neurofeedback approaches used such as theta/beta, SMR/theta and SCP [slow cortical potentials] neurofeedback nor a differential efficacy for the 3 domains' (Arns, Ridder, Strehl, Breteler, & Coenen, 2009).

In spite of this vagueness concerning the training, practitioners continuously state that their clients' brainwaves will be normalized by their practices, and that these effects can be made visible by means of (quantitative) EEGs (qEEGs). By comparing the EEG of a client with a database in which thousands of EEGs are compared and averaged, the deviances can be calculated and visualized in red or yellow (frequencies that are higher than the mean), blue (too low), or green (average) colors. Showing these qEEGs to the client in question can be described as quite literally a form of a normalizing practice (Foucault, 2004). The client sees at a glance if his or her brain frequencies deviate and is immediately offered a way to normalize these. The yellow, red, and blue spots in the brain map, one could say, ask to be changed into a green, normal brain (Fig. 6.1).

Changing yellow, red, or blue spots into green appears to be an 'objective and measurable method', as one of the clients expresses, but in interviews with practitioners, this relation between the qEEG and the training appears to be less clear. One practitioner firmly states: 'In 70 % of the cases I see "what you train is what you get".' According to this practitioner it also occurs that the improvement pervades the whole brain, instead of affecting only the trained spot: 'I train this person and the session data don't change that much, but then, when I take another EEG, you see the

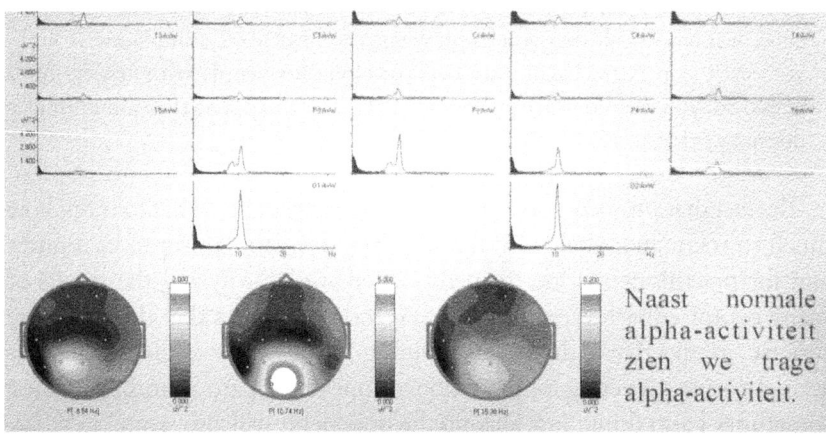

Naast normale alpha-activiteit zien we trage alpha-activiteit.

Fig. 6.1 Part of a qEEG report as presented to the client. The text reads: 'Apart from normal alpha activity, we can see low alpha activity' (Used with permission of Roland Verment)

whole brain being normalized' (5). Other puzzling performances concern improvements that appear at unexpected places. One practitioner talks about a case study that was demonstrated during a neurofeedback conference: '[The practitioner] trained a particular frequency at a particular site according to the brain map. (…) And the [problem] stopped, but the EEG…, the changes were somewhere else completely. Not where they were expected' (2). That is to say, the behavior of the qEEG during a neurofeedback therapy can be described as a 'puzzling material performance': when the practitioner trains one specific spot he or she has to wait and see if the brain will do what is expected, or will be normalized completely, change at a different spot, or does not change at all.[9]

The theory of neurofeedback is based on changing someone's brainwaves, but as demonstrated, there is no one-to-one relationship between the protocol and the complaints, or the training and the qEEG. Furthermore, it is somewhat vague who or what is performing at what moment. Some practitioners, for example, leave the whole process to the computer, and simply report: 'We are not physicians; the computer takes the unbalances out of the brain' (8). Another practitioner, who also works as a research assistant, after being asked about the training she gives to the eleven-year-old participants, explains: 'I know how to operate the machine, and how to hook them up and what instruction to begin with, but really in terms of background knowledge of alpha/theta you'd better talk to someone else' (2). And another practitioner brings up the computer as a material agent by explaining: 'I don't know the program well enough to say how it [the computer] sees what to correct for different people with different baselines' (6).

The practitioner is clearly not always the leading agent in the process: he or she struggles with protocols that do not always work, brains that can change unexpectedly, and computer programs that define what should be corrected. To phrase this in Pickering's words: 'human and material agency are reciprocally and emergently intertwined in a struggle'

---

[9] That is to say, *if* the expected changes are checked in the EEG. The supervisors of the neurofeedback course simply state: 'Nobody wants a pre and post q[EEG], because it doesn't do anything to give someone a paper and say "Look! You have been fixed"'. (3) Clients, however, object to this, for example by stating: 'If certain therapies or devices claim to change your EEG, than it *has* to change your EEG'. (12)

(Pickering, 1995, p. 21). As a result the neurofeedback practice actually becomes a process of 'trial-and-error tinkering', in which the practitioner has to find out which protocol, method, or frequency range works for which client. One practitioner plainly demonstrates how he more or less experiments on his clients:

> You activate, for example on the left. And you take a specific frequency range, and notice that the person becomes agitated, but still he has to be activated. Well, then you drop down those frequencies you are training, those that are connected with less activation, and see if you can reach your goal now. (7)

This process of experimenting on the client is a process of trial and error. Sometimes, this trial results in a serious error, but that does not seem to be a problem to the practitioners. One practitioner explains:

> [If that happens] I readjust it a little. It is a very subtle process. I had some-one with anxiety, and I put the alpha on 10–14 hertz. (...) And this person came back with a worsening of her complaints. Then I screwed it down again and at once it was all right. (5)

It is mostly the practitioner who decides to change the protocol after conversations with a client, but it can also be the client who calls the practitioner to account. As one practitioner clarifies: 'This client returned after two weeks and said: "Well, I don't know what happened, but this is not how it is supposed to be, because at home I kicked in the kitchen door." He had such a short fuse, and I had activated that' (7).

It also happens that the experimenting practitioner and the experiencing client closely work together during the process. One practitioner reports: 'One of my clients was giving good feedback, and she knew when I was doing it right, when I was doing the training with her. And she would say if she would feel less well, and then we would stop' (2). She stresses that in this particular case, the perception of the client is more important than the results on the computer screen: 'If you just do this with the computer: you can do too much left, too much down and then it doesn't look that right. So you need to pay attention to how the client is feeling.'

Neurofeedback appears to be an experimental practice. It is a process of tinkering (Knorr-Cetina, 1981; Pickering, 1995) between a practitioner who actively tunes his machine, and then passively stands back to watch what will happen with the client. This can be compared with how Pickering analyzes scientific practices:

> As active, intentional beings, scientists tentatively construct some new machine. Then they adopt a passive role, monitoring the performance of the machine to see whatever capture of material agency it might effect. Symmetrically, this period of human passivity is the period in which material agency actively manifests itself. Does the machine perform as intended? Has an intended capture of agency been effected? Typically the answer is no, in which case the response is another reversal of roles: human agency is once more active in a revision of modeling vectors, followed by another bout of human passivity and material performance, and so on. (Pickering, 1995, pp. 21, 22)

In the case of neurofeedback, it might seem that the practitioner is the intentional being, in the sense that he or she decides on the protocol, tunes the machine, waits for a reaction in the client, and makes some adjustments. However, it could also be the computer that decides on the protocol (e.g. by seeing what to correct), and instead of the client it can also be the EEG that responds (change at an unexpected place). Other agents can also become 'temporally emergent', like deviant personality types (no A, B, or C, but a mixture), feelings (experiencing a low stimulation as too much), or brains (the complaints do not fit the EEG). That is to say, although there is obviously a dance of agency going on, its choreography remains unclear, which makes it very complicated, or even impossible, to predict what will happen next.

Collecting the Results

In Pickering's description of scientific practices the choreography of the open-ended dance of agency can become 'relatively fixed' (Pickering, 1995, p. 102), and this is where an important difference between constructing quarks and selves appears. Hunting for quarks and changing brainwaves can be compared in the sense that humans and non-humans together perform a dance of agency, but the comparison falls short concerning the end point. After all, the purpose of a therapy is not a changed brain, but a changed client. Constructing a quark is a 'temporal' process which means that it has an end point somewhere, that is, when the material quark and the concept of a quark are 'interactively stabilized' (Pickering, 1995, pp. 17, 83). The moment when the effects of neurofeedback on the subject are finished or elaborated, on the other hand, usually remains unclear.

This instability of the result is demonstrated by the efforts practitioners make to help their clients recognize what is improved in their brains, feelings, performances, or lives. Practitioners appeal to many actors, from parents, brain maps, to specific results which make the client conscious of his or her changed state. One practitioner explains:

> If you ask 'do you notice anything?' they mostly answer 'no'. And if you keep on asking about school, work, or whatever, they often say 'oh, yes. I do concentrate better'. Or they feel more relaxed, or less aggressive, or have a better outlook, or their marks are improved, or whatever. (...) It is a specific way of asking to find this out, because they do notice something; of course they notice something. And the environment also is very important, especially parents. Or you hear that teachers have said that a child can stand more, or has become more social, or less aggressive. (6)

This practitioner shows that the improvement can actually be found everywhere: in school or work performances, clients' psychological well-being, or their behavior toward others, and can be reflected by various actors such as parents, teachers, or grades. It is up to the practitioner to identify these changes, for example, by employing a specific way of asking, so that the client can recognize them too. To make sure that the expected change is not missed, some practitioners start every session with asking what has been improved in their client since the last training.

Practitioners give several reasons to explain why it can be hard for clients to recognize the effects of neurofeedback. One researcher clarifies: 'The improvement is so gradual that people don't see it is the neurofeedback that has changed them. They see the change as normal, as having a good day' (1). According to a practitioner, it can also occur that clients do not notice they are cured from some complaints because they only focus on those problems they still have. Furthermore, it can be the case that the change takes some time before it sets in, as demonstrated by this practitioner talking about a boy whose parents let him quit after 20 sessions, because they hardly noticed an effect: 'Two months later I sent them an e-mail to see how he was doing. And he had quit his medication, the antipsychotics, and his ADHD-medication was halved. (…) Well, for neurofeedback this is quite a good effect' (5).

This construction, or collecting, of the changes is also demonstrated when the same practitioner talks about the recovery of another client:

> The first symptom was that he took a book from the bookshelf. His parents thought this was really strange. A couple of sessions, nothing happened but then he started to read texts on trucks. It appeared that his reading level had reached the average for his age. (…) and in fact, I think his IQ is now about 80 or something. And his medication has been reduced. (5)

There are many cases and examples of clients, parents, and practitioners who state that neurofeedback cures or improves the client. Designated successes often concern skills and performances, but sometimes the client's behavior in specific situations can be a sign of improvement too. One practitioner gives an example of a man who wanted to go to the casino: 'But when he stood in front of his motorbike he thought "no, I will not go". He thought this was because of the neurofeedback' (6). Furthermore, he explicates that changes cannot only be signaled in someone's behavior, performances, or development, they can also be psychological:

> People have told me that it reduces their stress, they experience some space in their head and the capability to distinguish main and side issues. If everything rushes upon them they can distance themselves from their worries. And, I also think it can give more energy, this neurofeedback. This will help too, that you can increase someone's strength. (6)

Besides behavioral and psychological results, physical effects are sometimes also experienced, such as 'a tingling in the head', 'a deep relaxed feeling', 'headache', 'a feeling of being drugged', 'a euphoric feeling', 'a feeling of space in the head', and someone even talks about the feeling of 'experiencing a cerebral haemorrhage'.

Recognizing and collecting the results of a neurofeedback therapy seems to be a complex practice. According to the accounts of the practitioners, effects can appear inside the client or outside in the world. Sometimes, they are recognized spontaneously by the client, but more often, changes are brought up with the help of the interviewing techniques of the practitioner, or by external actors who notice and determine the improvement. It also occurs that clients (e.g. teenagers) state they are cured, while others (e.g. parents) decide they are not ready yet. Effects are collected from everywhere: from improved results in school or work, better psychical well-being, physical experiences, more social contacts, a reduction of medication, or in a decision not to go the casino. According to practitioners, progress can appear at every moment: sometimes it only takes 1 session, sometimes it requires 80 sessions, and sometimes the change starts weeks after the training has stopped.

The success of neurofeedback is quite unpredictable, and actually depends on an agreement between the practitioner and the client concerning many other actors. However, as demonstrated in Chap. 4, listening to users of neurofeedback brings another effect to the fore. Some of the actors that emerge during the neurofeedback process also appear in the descriptions of people doing neurofeedback, but as entities of the self. Hence, I argue that some of the actors that emerge during the dance of agency are not only 'temporally emergent' during the process, but they become part of the product: a new self for the client.

The Self as a Dance of Agency

In the process of neurofeedback many actors are involved, of which some become part of the client's self-conception. Practitioners teach their clients that their problems are brain problems, help them to perform the neurofeedback training, and to recognize the results. They ease their client's psyches, correct their bodies, and teach them the ins and outs of their brainwaves, peaks, spots, explosions, and other brain entities. While doing this, they sometimes refer to the computer as the expert (the computer sees), the brain as an actor (the brain knows), brain entities as the cause of the problems and as responding to the training. Neurofeedback is often compared with doing meditation, sometimes it is combined with psychotherapy, and various actors—from parents to test results—are involved in helping the client recognize the results. In this whole process it obviously is the client him- or herself who is addressed; neurofeedback is called a method of self-confirmation, self-control, or self-enhancement. The aim of this process is a better self, a better life, a better functioning for the client.

However, as described in Chap. 4, the result of this process is not only (nor always) a better self, but also a much more complex self. Neurofeedback users suggest that their selves are, or are in, their brains, but to do neurofeedback they have to make a distinction between their selves (I) and their brains (it). They take over the language of their practitioners and explain their selves, problems, and lives with entities like spots, neurons, and waves; combine these with computer metaphors and spiritual practices; retain the traditional souls and psyches; struggle with earlier biological, evolutionary, or psychological self-conceptions; and create a self that is probably best understood by getting friends and family members involved in the neurofeedback practice.

That is to say, describing neurofeedback as a technology of the self or describing neurofeedback as a dance of agency conjures up the same kinds of entities working upon the self of the client. And although it is hard to define how freely users constitute this new way of self—since users are obviously steered and controlled by all kinds of other actors—every user constitutes his or her own mix of entities, combines it with his or her own ideas and experiences, and passes it on to his or her own friends and families. In other words, the neurofeedback process can be described as a dance of agency that constitutes an extended mode of self, which is again a dance of agency.

Conclusion

Pickering proposes a shift from an epistemological to an ontological way of thinking about science. According to Pickering, the world performs, and human and non-human are linked together in dances of agency. Examining these dances closely will demonstrate that the material world is 'not fixed, static and knowable, but endlessly lively' (Pickering, 2009). In this chapter, I followed Pickering's idea and described a clinical practice as a dance of agency, or, to use one of Pickering's other fascinating concepts, as ontological theater. In my exposition of the practice, neurofeedback is not just a simple act between two human agencies and a computer, but a play in which many collaborating and competing actors are involved and together perform a dance in which it is not completely clear which actor is in charge.

I described several necessary steps to make neurofeedback a success. First, there are many actors working upon the potential client to turn him or her into a cooperative client. Parents and psychiatrists help the person recognize his or her problems, brain awareness campaigns (leaflets, Internet sites, newsletters) turn these problems into brain problems, and neurofeedback specialists offer a solution by demonstrating their practices in the media. Next, actors ranging from practitioners, Neury Bears, to footstools are involved to motivate the client's mind, body, and brain. Protocols, individual brains, computer programs, and EEGs have to work together in a process of trial and error, and the expected change has to be recognized and pointed out, somewhere inside the client, or outside in the world. The result of this, I argued, is again a dance of agency, but now concerning the self of the client. Several of the actors that 'temporally emerged' during the neurofeedback process—brainwaves, computers, colored spots—keep on working on the self of the client.

To phrase my conclusion in Pickering's terms: human and non-human actors are linked together in a dance of agency and examining this dance conjures up an image of the world as endlessly lively. The result of this dance is a new being in the world. While it is not very obvious if and how the client is cured, restored, or enhanced, a new kind of self—one that has been extended with all kinds of entities that emerged during the process—has clearly been brought into being.

References

Arns, M., de Ridder, S., Strehl, U., Breteler, M., & Coenen, A. (2009). Efficacy of neurofeedback treatment in ADHD: The effects on inattention, impulsivity and hyperactivity: A meta-analysis. *Clinical EEG and Neuroscience, 40*(3), 180–189. doi:10.1177/155005940904000311.

Barad, K. (2003). Posthumanist performativity: Toward an understanding of how matter comes to matter. *Signs: Journal of Women in Culture and Society, 28*(3), 801–831.

Callon, M. (2007). What does it mean to say that economics is performative? In D. A. MacKenzie, F. Muniesa, & L. Siu (Eds.), *Do economists make markets?: On the performativity of economics*. Princeton, NJ: Princeton University Press.

Collins, H. M. (2000). What is tacit knowledge? In T. R. Schatzki, K. K. Cetina, & E. von. Savigny (Eds.), *The practice turn in contemporary theory* (pp. 115–128). London, New York: Routledge.

Dehue, T. (2008). *De depressie-epidemie*. Amsterdam: Uitgeverij Augustus.

Dumit, J. (2003). Is it me or my brain? Depression and neuroscientific facts. *Journal of Medical Humanities, 24*(1), 35–47. doi:10.1023/A:1021353631347.

Foucault, M. (2004). In M. Bertani, A. Fontana, F. Ewald, & D. Macey (Eds.), *"Society Must Be Defended": Lectures at the Collège de France, 1975–1976*. London/New York: Penguin.

Hacking, I. (2006). Making up people. *London Review of Books, 28*(16), 23–26.

Hacking, I. (2007). Kinds of people: Moving targets. *Proceedings of the British Academy, 151*, 285–318.

Kendall, G., & Michael, M. (2001). Order and disorder: Time, techology and the self. *Culture Machine*. Retrieved from http://www.culturemachine.net/index.php/cm/article/view/242/223

Knorr-Cetina, K. (1981). *The manufacture of knowledge: An essay on the constructivist and contextual nature of science*. Oxford: Pergamon Press.

Latour, B., & Woolgar, S. (1979). *Laboratory life: The construction of scientific facts*. Princeton, NJ: Princeton University Press.

Pickering, A. (1995). *The mangle of practice: Time, agency, and science*. Chicago/London: University of Chicago Press.

Pickering, A. (2007). Ontological theatre. Gordon Pask, cybernetics, and the arts. *Cybernetics and Human Knowing, 14*(4), 43–57.

Pickering, A. (2009). Science, contingency and ontology. Retrieved from http://hdl.handle.net/10036/81575

Pickering, A. (2010). *The cybernetic brain. Sketches of another future*. Chicago/London: The University of Chicago Press.

Polanyi, M. (1962). Tacit knowing: Its bearing on some problems of philosophy. *Reviews of Modern Physics, 34*, 601–615.

Rabinow, P., & Rose, N. (2006). Biopower today. *BioSocieties, 1*(02), 195–217. doi:10.1017/S1745855206040014.

Rose, N. (2007). *The politics of life itself: Biomedicine, power, and subjectivity in the twenty-first century*. Princeton, NJ: Princeton University Press.

Roy, S. C. (2008). 'Taking charge of your health': Discourses of responsibiltity in English-Canadian women's magazines. *Sociology of Health and Illness, 30*(3), 463–477.

7

Reflection and Conclusion

How did we become neurochemical selves? How did we come to think about our sadness as a condition called 'depression' caused by a chemical imbalance in the brain and amenable to treatment by drugs that would 'rebalance' these chemicals? How did we come to experience our worries at home and at work as 'generalized anxiety disorder' also caused by a chemical imbalance which can be corrected by drugs? (Rose, 2003, p. 46)

According to Nikolas Rose, we have become neurochemical selves, and in the article this quote stems from his analysis of how this occurred. The statement that we have become neurochemical selves, however, has recently been criticized. Some scholars, studying the impact of neuroscience on the social realm, argue that Rose's arguments are 'overstretched' or even that it is partly through the sociological gaze itself that neurological subjectivity is constituted (Bröer & Heerings, 2012; Pickersgill, Cunningham-Burley, & Martin, 2011). In light of this criticism,[1] this chapter reflects on my

---

[1] With which I do not fully agree, since I have never interpreted Rose's statement as strictly as these authors take his words. Of course, people are not 'only' neurochemical selves, but neuroscientific explanations are clearly present in everyday language.

141
J. Brenninkmeijer, *Neurotechnologies of the Self*,
DOI 10.1057/978-1-137-53386-9_7

claim that neurofeedback users constitute an 'extended self': a self that is extended with a brain, and forms an assemblage of various neurological, psychological, biological, social, mechanical, spiritual, and other entities.

Sociologists, anthropologists, and psychologists have studied how neuroscience—in the form of pills, scanners, therapies, self-help books—has influenced the self. The results of these studies are twofold: it is obvious that using brain technologies influences the self, but it is also obvious that the self is more than only the brain. According to Rose, personhood is no longer concerned with the mind or the psyche, but with the brain. This new knowledge makes that we take ourselves to be different kinds of persons, and can be understood as a shift in human ontology: 'It entails a new way of seeing, judging, and acting upon human normality and abnormality. It enables us to be governed in new ways. And it enables us to govern ourselves differently' (Rose, 2007, p. 192). Other scholars also described this impact of brain knowledge on the self. Joseph Dumit, for example, studied how brain images can alter people's understanding of their own body, and uses the concepts 'objective-self' and 'pharmaceutical self' (Dumit, 2003, 2004). Fernando Vidal performed a historical study on 'brainhood', with which he referred to the 'quality of being a brain' (Vidal, 2009), and Davi Johnson Thornton analyzed the impact of the message that 'you are your brain' (Thornton, 2011). Other scholars, however, nuance this impact of the brain and emphasize that people who are confronted with their neurological constitution do not simply become neurologic subjects but use the heterogeneous language of psychological and physiological statements (Bröer & Heerings, 2012; Choudhury, McKinney, & Merten, 2012; Martin, 2010; Pickersgill et al., 2011).

This book contributes to these studies regarding the effects of neuroscience by exploring how working on the self with a brain device changes people's concepts of themselves. I relied on Foucault's historical explorations of 'technologies of the self' and presented a historical, ethnographic, and theoretical analysis of the mode of subjectivity that is constituted when people use a brain device to change themselves. My ethnographic findings of neurofeedback users indicate that people who work on themselves with a brain device extend their selves with a brain and several other physiological, psychological, technical entities that emerged in the process. Hence, in one way or the other, this conclusion agrees with the studies describing the impact of the brain, and with those emphasizing the heterogeneity of explanations regarding the self. Moreover, my conclusions do not only confirm but also elaborate on those studies, since neurofeedback users exhibit that it is not only neurological and psychological explanations, but also the technical process and the spiritual connotations that contribute to the constitution of a new way of being oneself.

However, Pickersgill et al. also argue that it is, in part, 'the sociological gaze itself' that constitutes neurologic subjectivity. They do not claim that sociologists influence the way people perceive themselves, but that sociologists' ideas of people's selves are not completely correct. They explain their argument, for example, by showing that individuals also resist a neurologic or reductionist figuration of subjectivity, and that even neuroscientists expressed surprise at this claim that we are now 'neurochemical selves' (Pickersgill et al., 2011, pp. 354/355). Although one might wonder if these respondents would have accepted the conclusion of Pickersgill and colleagues that they could best be understood as bricoleurs, 'piecing together diverse knowledges concerning psyche, soma and society' (Pickersgill et al., 2011, p. 361), it makes sense to discuss the role scholars might have in the constitution of their findings. Since I performed this research, by creating questions and choosing the interviewees and practices to observe, by selecting people's phrases and connecting these to 'others' phrases', and by combining the results with theoretical and philosophical statements of authors I prefer, one might wonder if I am not also the one who created this kind of self, and what use this might have. That is to say, instead of asking how neurofeedback subjects became extended selves, it also makes sense to wonder if this 'extended self' is not simply an artifact of my methods of research. In the following, I reflect on what preceded and discuss in what way using a brain device constitutes a new mode of being oneself, and why it is important to make these processes visible.

Does It Work?

Just like Pickersgill's neuroscientists did not recognize themselves in Rose's analysis of neurochemical selves, most neurofeedback users will, probably, not see themselves as human–machine cyborgs, or as selves that are extended with autonomous brain entities. According to my subjects, doing neurofeedback will result in a different way of being a self: a self they want to be, accept, or control. That is, a 'me without my problems'. Even if they want, and can, follow my arguments about the 'extended self' which emerges in their striving for a better self, they do not simply agree. I discussed my results with one of my interviewees who responded: 'I don't think there is a difference between the brain and the self. In my opinion, the self is an illusion, a product of the brain' (12).[2] Moreover, some practitioners and other people responding to my work do not even see the relevance of my explanations of the self. During meetings and conversations, practitioners generally tried to convince me that neurofeedback really does work, and especially ethicists and psychology students often wondered why I did not take a more normative stance and explained that neurofeedback does not work.

In other words, when neurofeedback users claim that they have become themselves by doing neurofeedback they do not mean that they have become extended selves, but that the therapy helped to solve their problems. As a result of such claims, many other people would like to know if brain devices 'work' in the sense that their supposed therapeutic effects are scientifically credible. Chap. 2 formulated an answer to this question by explaining that the scientific credibility of brain devices partly depends on how they are demonstrated. Following Ashmore, Brown, and Macmillan (2005), I argued that therapeutic effects are difficult to demonstrate since they concern the mind: a topic that is not directly observable but also ubiquitous. I analyzed the histories and contemporary uses of several brain devices regarding three modes of demonstration that psychologists employ to convince others of the reliability of a phenomenon (Ashmore et al., 2005). In sum, I concluded that brain devices are better demonstrated in a personal, self-experimenting setting than in a scientific or polemic domain.

However, users of neurofeedback and other neurotechnologies of the self illustrate that giving insights in the question of scientific credibility

---

[2] At another moment, however, I asked him if he thought he was his brain and he responded: 'There is also one part of me that doesn't want to see it at all in that way. And [...that part...] refuses to see myself as nothing more than a bio-organic robot.' See Chap. 4.

is not sufficient to find out if a device works or not.[3] Many people—scientists, practitioners, clients, self-experimenters—try out these devices on themselves or others to find out if they can produce an effect, and hence they demonstrate the effects of these devices in various ways. In spite of elaborating on discussions concerning the scientific validity, this book argued that these devices have effects, as long as they are used.

Neurofeedback Tribe

To find out which effects brain technologies can have on their users, Chap. 4 analyzed how neurofeedback users think about, act on, and constitute them-/their selves. My research showed that to do neurofeedback, people make a clear distinction between their selves and their brains, and bring various other psychological, biological, technological, and sometimes spiritual entities into being that start working upon their selves, lives, and histories. That is to say, I argued that doing neurofeedback—at least for some people—constitutes a new way of being oneself.

However, this analysis raises the question again whether one can say something about the self that many people do not consider as relevant, or as reality. Such disagreements about relevance and reality often occur in response to scholars who study phenomena from a symmetrical point of view, in which they do not distinguish 'true' and 'false' scientific statements, experts or lay people, or human or non-human behaviors (Latour, 1987, 1999; Sismondo, 2011). One way out of these disagreements about relevance or reality is to draw a comparison with an anthropologist who studies the rituals and beliefs of a tribe.[4] Just as it makes no sense to ask an anthropologist if the beliefs of the tribespeople are true or false, or good or bad, or if their fetishes (e.g. a holy tree or a statue of the Virgin Mary) can or cannot act, it is also out of place to ask scholars who study science with a symmetrical approach to decide about the scientific value or ethical norms of the techniques they study.

---

[3] To give a different example, antidepressants are generally seen as scientifically credible while many scientists discuss their efficacy (e.g. Ioannidis, 2008).

[4] This kind of anthropological approach was introduced by Bruno Latour and Steve Woolgar who studied a research laboratory setting as a tribe (Latour & Woolgar, 1979).

To study the neurofeedback 'tribe' from a symmetrical or anthropological point of view, I abandoned the boundaries of categories that are normally carefully separated, like psychological, physiological, and mechanical explanations, intentional and unintentional beings, and scientific and common-sense statements. It was not my intention to make things more complicated with this symmetrical approach, or this lumping together of categories; rather, I wanted to take seriously what users actually do and say. Ignoring the totally yellow spot, the controlling neurotransmitters, or the self-resetting system, and sending these to the realm of 'confusion', or deciding that they are less true than psychological states or categorized diagnoses, would disregard the experience of the users. Neurofeedback users bring up various unusual and sometimes abstract entities, which have an impact on their lives, relationships, and decisions, so although it might be uncommon to take these entities seriously, it would be negligent not to take them into account. Moreover, while most neurofeedback users will not literally see their selves as extended with multiple entities, my analysis of their explanations and activities can make them aware of the amalgamations they bring into being.

Emergence of the Extended Self

Studying neurofeedback as a technology of the self, however, is only one part of the story. To understand how neurofeedback users constitute a new way of being, it was also necessary to analyze how other actors helped creating this mode of self. Since neurofeedback is a therapy between humans, but also a technology performed by computers and electroencephalograms (EEGs), it was important to include non-humans in my account of neurofeedback, but without making humans and non-humans equivalent. Hence, I analyzed the neurofeedback process with the work of Pickering, who describes scientific practices as dances of agency between human and material entities. This analysis made clear that those entities working upon the self of the user also have a vital role in the neurofeedback process. That is, the neurofeedback process is a dance of agency between a client, a computer, and a practitioner, but also involves struggles between physiological and psychological entities, computer programs and EEGs, and material and spiritual entities—and a comparable assemblage of entities can be encountered in the ways users constitute themselves.

Moreover, as demonstrated in Chap. 3, these entanglements of brains and selves, humans and machines, and material and spiritual beliefs already emerged in the work of four central figures in the history of neurofeedback. I described how Hans Berger connected brainwaves to mental activity, but also demonstrated its difficulty by trying to explain its interaction. As a result, it is actually unclear whether Berger should be considered as a monistic, a dualistic, or maybe even a holistic thinker. This disagreement, which can also be traced in statements and acts of contemporary neurofeedback users who on the one hand try to improve themselves by working on their brain and simultaneously distinguish themselves from their brain, demonstrates that the line between monism and dualism is not as clear as is often claimed.

Moreover, one of Berger's followers, William Grey Walter gave rise to a deterministic brain with his alpha theories, but with his theta experiments, he conjured up the image of a controllable brain. That is, in his work a distinction between the subject (or the self) and the brain can be traced which resonates with statements of contemporary neurofeedback users, who, for example, explain in Chap. 4: 'It seems that at the moment you start focusing, your brain interrupts (…).' Furthermore, the connection, and with this the entanglement, between human and machine that was made by Walter, can also be found in the statements of contemporary neurofeedback users. Phrases like a 'defragmentation of your computer', 'a computer wiring me', 'my system is unstable' strongly resemble Walter's claim that 'the brain has a capacity for resetting itself, for setting up its own wiring' (Walter, 1953, p. 141). Also, in the work of the psychologists Joe Kamiya and Barry Sterman, as well as in the explanations of contemporary neurofeedback experts, the steering brain and the modulating (spiritual) self emerged and maintained a complicated relationship.

That is to say, the self was connected to the brain in the work of Berger, replaced by the brain in the work of Walter, and restored as an autonomous entity that could act upon the brain in the work of Kamiya and Sterman. In these developments, brain-related entities were distinguished, material and spiritual ideas became entangled, human–machine connections arose, and a complicated relationship between the brain and the self emerged in which they started to control each other. In the explanations of contemporary neurofeedback users and practitioners, these entities and entanglements are also clearly visible. Contemporary neurofeedback users, for example, talk about their 'bad beta' or their self-resetting system. They state that they use their 'will power' to control their brainwaves, explain that they give their brain 'an assignment', or explicitly try not to 'pay attention' to its actions. The recurring frustration concerning the mind–brain relation is also revealed in the statements of contemporary neurofeedback users; for example, when they say 'I had always thought that I was controlled by myself [instead of by neurons]' or '[the feeling part of my brain] refuses to see myself as nothing more than a bio-organic robot'.

Doing neurofeedback creates several new ways of perceiving oneself, and with this new ways of acting and being oneself. This self is not universal or fixed, but analogies between users' statements can be traced. Neurofeedback users split their selves from their brains, refer to autonomous brain (map) entities, compare themselves with machines, and maintain a commitment to the traditional souls, minds, and psyches. As Chap. 3 showed, these struggles between selves and brains, humans and machines, spiritual and material entities, and psychological and physiological explanations should not be considered as confusion made by lay people not being able to cope with a new ('modern') scientific way of thinking, but are a result of a historical quest to grasp the (spiritual) self with a brain device. That is, these assemblages derive from the activity to improve or understand the self by working on the brain, and hence form an argument against the reductionist view of the self.

Whose Self, What Self

By taking seriously what people do and say I gave a different explanation of the working and effect of neurofeedback than most people might expect. Instead of adopting or rejecting the brainwave theories and brain-based improvements, I showed that the reductionist vision of the self—as promoted by neurofeedback practitioners as well as many other scientists—is not contained in people's explanations about themselves. Moreover, working on the brain to improve the self does not reduce the self to the brain, but extends the self with a brain and all kind of other entities emerging in the process.

This might seem a complicated argument since it disagrees with most scientific as well as individual experiences. The self is one of the most difficult and ambiguous concepts that exists in psychology, philosophy, and neuroscience, and there are continuing controversies about its existence, substance, and location. At the same time, however, to most people the self is as plain and as present as the nose on their face, and when brought up in conversations there is no confusion about the concept at all. This clearness of the self in everyday life is actually quite amazing since the meaning of 'self' shifts per person, time, situation, and also per scientific discipline. Even within one person, at one specific situation and moment, the self can be something that is not present (I am not myself), unknown (I don't know myself), or unmanageable (I am not in control of myself). Moreover, most of my interviewees stated that they were not (or had never been) themselves but needed the neurofeedback to become themselves.

Such statements in which the person who says 'I' is not the same as the self illustrate that the agent of this self changes: the 'I' sayer declares that he or she is not the controller of his or her behavior. The agent at this moment can, for example, be a strong emotion (I was so angry, I was beside myself), a disorder (that was not me, that was my ADHD), or a technology (I was not myself, because my smartphone, hearing aid, and so on, did not work properly). These examples do not only indicate that the self *relates* to many entities; the self is *constituted* by these entities since the loss of an important entity gives the feeling of being out of control. In other words, the agency of the self is often not only in the person (or in the brain), but largely in the external(ized) entities. This is why the philosopher of mind Andy Clark argues:

> There is *no self*, if by self we mean some central cognitive essence that makes me who and what I am. In its place there is just the 'soft self': a rough-and-tumble, control-sharing coalition of processes—some neural,

some bodily, some technological—and an ongoing drive to tell a story, to paint a picture in which 'I' am the central player. (Clark, 2003, p. 138, see also 2008)

Clark's view is interesting and resolves many difficulties concerning the various entities—technologies, people, objects, brains—that give form and meaning to people's private selves. However, where Clark describes the self as a 'control sharing coalition of processes' the neuro-feedback users showed there is not always peace and quiet in this coalition that forms the self. Especially when people are not satisfied with themselves they start working on their selves and in this process they mobilize many entities like brainwaves, will power, emotions, or yellow spots that start struggling around. In this process of working on the self many internal and external entities appear that together constitute this coalition of the self.

Slowing Down the Activity

Some scholars describe the amalgamation of immaterial and material entities as a form of post- or transhumanism, but according to scholars like Clark or Latour, coalitions between subjects and objects—as one might call them—are nothing new or modern (Clark, 2003; Latour, 1993). Moreover, Latour argues that while in 'pre-modern' cultures hybrids (for instance, holy trees) are widely accepted, the attempt of modern cultures to distinguish subjects from objects brings hybrids into being even more quickly: speed bumps that help you drive safely, smartphone that remember your appointments, but also atomic bombs that threaten nations. He concludes that exposing the hybridization practice will slow down this process, and adds: 'It is from this slowing down, from this moderation, from this regulation, that we await our morality' (Latour, 1993, p. 142).[5]

---

[5] I used the translation of Hans Harbers (1995) who revealed a mistake in the English translation of Latour (1993). According to the English edition this citation reads: 'This slowing down [...] is what we expect from our morality', while the original French sentence is 'C'est de ce ralentissement [etc.] que nous attendons notres moralites.'

Latour's argumentation helps to clarify why it is useful to take seriously what neurofeedback users do and say, and how they constitute a new mode of self. While neurofeedback clients are told that their selves and their problems can be improved by changing their brains with a computer, they extend their selves with all kinds of brain, brain map, computer, and other entities. Stated more generally, while scientists try to clarify the human subject by objectification (e.g. in waves, peaks, neurons, or spots), subjects start to extend their selves with these objectified parts. As a result, it becomes increasingly complicated to understand who or what (me, my system, neurons, yellow spot) is in charge in/of the self. Although some practitioners and clients claimed that neurofeedback makes people more themselves and allows them to take responsibility, users also expressed feelings of loss or fear, since they felt controlled by their neurons, brainwaves, or the computer. Moreover, Chaps. 4 and 6 demonstrated that the performing agent (in the process and the self) dances around, which makes it difficult to claim that the self is or becomes an autonomous and responsible entity by doing neurofeedback. This inconsistency between the promise of neurofeedback (becoming oneself) and my findings (extending oneself) requires some clarifications about notions of autonomy, responsibility, and freedom.

In his book *Moralizing Technology* (2011) the philosopher of technology Peter-Paul Verbeek explains that the notion of autonomy has become highly problematic in technological societies. After all, many of our actions and decisions are technologically mediated, for example, by neurotechnologies, ultrasound scans, or speed bumps. Moreover, Verbeek argues that we cannot hold on this notion of human autonomy since in our technological culture 'humans and technologies do not have separate existences anymore' (2011, p. 16). However, this does not mean that people cannot be responsible or free. On the contrary, he argues, if we deal with technological influences we can become free: freedom becomes an *activity*, a practice of dealing with (technological and other) power (2011, p. 73). That is to say, in order to become free and responsible subjects, Verbeek argues that we should recognize the constitutive role of technology in human existence.

Latour and Verbeek both analyze how humans and non-humans hybridize in modern or technological societies, and both argue that it is necessary to recognize this process to deal with it. Following Latour and Verbeek, I would like to argue that—irrespective of whether brain devices are considered scientifically credible or not, and of whether subjects recognize themselves as extended selves or not—it is important to pay attention to the constituting effect of neurotechnologies on the users' subjectivity. Making people aware of the ways they talk about and act upon themselves gives them the opportunity to deal with the amalgamations they constitute. That is to say, if we want to slow down or regulate this process of extending the self, for example, because it makes the self increasingly complicated to understand, feel free about, or responsible for, it is worthwhile to start making it visible.

Conclusion

Rose wanted to understand how people became neurochemical selves, and his critics countered that we are not 'only' neurochemical selves, and, moreover, that this neurologic subjectivity might (partly) be an artifact of the sociological gaze itself. In this chapter I used these arguments to reflect on my analysis that neurofeedback users constitute an extended self. Although neurofeedback users sometimes claimed that they are their brain, that they are primates that can be trained, or just miss a certain substance in their brain (statements that all resemble Rose's analysis that we have become neurochemical selves), they also accentuated their selves, lives, feelings, and psyches (which agrees with statements of Rose's critics) and combined these with computer metaphors and spiritual practices. That is, to a certain extent my analysis agrees with, and elaborates on, both Rose's claim and that of his critics.

However, that does not alter the circumstance that many neurofeedback clients, practitioners, and scientists may not see my analysis as relevant or as true. To put this discrepancy into perspective, I analyzed the scientific credibility of brain devices as a result of their scientific demonstration, and I described my position as an anthropologist studying the neurofeedback tribe. Hence, I argued that such discussions concerning reality and relevance are actually out of place, and might distract from the effects brain technologies do have on their users' subjectivity.

This book argued that using a brain device to understand or improve the self is a very complicated process (a dance of agency) in which material and non-material entities struggle around for control. Especially in those situations when people try to identify the self with the brain, this struggle seems to occur: neurofeedback clients started to emphasize their lives and psyches, neurofeedback practitioners become confused, and scientists become frustrated. In these processes, concepts like autonomy, responsibility, and freedom are used in multiple ways: people become themselves and lose control, take responsibility, and blame their brain, and combine their free will with their bad brain habits.

Returning to Foucault's concept of self might clarify why these contradictions, struggles, and misunderstandings occur when people try to grasp the self by working on the brain. In *Technologies of the Self*, Foucault writes:

> When you take care of the body, you do not take care of the self. The self is not clothing, tools, or possessions; it is to be found in the principle that uses these tools, a principle not of the body but of the soul. You have to worry about your soul—that is the principal activity of caring for yourself. The care of the self is the care of the activity and not the care of the soul—as substance. (Foucault, 1997, pp. 230–231)[6]

In this book I have demonstrated that care of the self is indeed more than care of the brain: the self is an activity that can be extended with, but not reduced to, a substance. When people worry about their brainwaves, neurons, self-resetting systems, or yellow spots, instead of their selves as an activity or process, they constitute an increasingly complicated way of being oneself. This self is complicated, not because it is an activity of multiple entities, but because it contradicts common understandings of being a self.

---

[6] Foucault discusses Plato's *Alcibiades I*.

References

Ashmore, M., Brown, S. D., & Macmillan, K. (2005). Lost in the mall with Mesmer and Wundt: Demarcations and demonstrations in the psychologies. *Science, Technology and Human Values, 30*(1), 76–110. doi:10.1177/016224 3904270716.

Bröer, C., & Heerings, M. (2012). Neurobiology in public and private discourse: The case of adults with ADHD. *Sociology of Health and Illness, 35*(1), 49–65. doi:10.1111/j.1467-9566.2012.01477.x.

Choudhury, S., McKinney, K. A., & Merten, M. (2012). Rebelling against the brain: Public engagement with the "neurological adolescent". *Social Science and Medicine, 74*(4), 565–573. doi:10.1016/j.socscimed.2011.10.029.

Clark, A. (2003). *Natural-Born Cyborgs: Minds, Technologies, and the Future of Human Intelligence.* Oxford: Oxford University Press.

Clark, A. (2008). *Supersizing the mind. Embodiment, action, and cognitive extension.* Oxford: Oxford University Press.

Dumit, J. (2003). Is it me or my brain? Depression and neuroscientific facts. *Journal of Medical Humanities, 24*(1), 35–47. doi:10.1023/A:1021353631347.

Dumit, J. (2004). *Picturing personhood: Brain scans and biomedical identity* (Information series). Princeton, NJ/Oxford: Princeton University Press.

Foucault, M. (1997). Technologies of the self. In P. Rabinow (Ed.), *Ethics: Subjectivity & truth, Michel Foucault on truth, beauty, & power 1954–1984* (pp. 223–251). London/New York: Penguin.

Harbers, H. (1995). Review of we have never been modern by Bruno Latour. *Science, Technology and Human Values, 20*(2), 270–275.

Ioannidis, J. P. (2008). Effectiveness of antidepressants: An evidence myth constructed from a thousand randomized trials? *Philosophy, Ethics, and Humanities in Medicine, 3*, 14. doi:10.1186/1747-5341-3-14.

Latour, B. (1987). *Science in action: How to follow scientists and engineers through society.* Cambridge, MA: Harvard University Press.

Latour, B. (1993). *We have never been modern.* New York/London: Harvard University Press.

Latour, B. (1999). *Pandora's hope: Essays on the reality of science studies.* Cambridge, MA: Harvard University Press.

Latour, B., & Woolgar, S. (1979). *Laboratory life: The construction of scientific facts.* Princeton, NJ: Princeton University Press.

Martin, E. (2010). Self-making and the brain. *Subjectivity, 3*, 366–381. doi:10.1057/sub.2010.23.

Pickersgill, M., Cunningham-Burley, S., & Martin, P. (2011). Constituting neurologic subjects: Neuroscience, subjectivity and the mundane significance of the brain. *Subjectivity, 4*, 346–365. doi:10.1057/sub.2011.10.

Rose, N. (2003). Neurochemical selves. *Society, 41*(1), 46–59.

Rose, N. (2007). *The politics of life itself: Biomedicine, power, and subjectivity in the twenty-first century.* Princeton, NJ: Princeton University Press.

Sismondo, S. (2011). *An introduction to science and technology studies.* New York, NY: John Wiley & Sons.

Thornton, D. J. (2011). *Brain culture: Neuroscience and popular media.* New Brunswick, NJ: Rutgers University Press.

Verbeek, P.-P. (2011). *Moralizing technology: Understanding and designing the morality of things.* Chicago/London: University of Chicago Press.

Vidal, F. (2009). Brainhood, anthropological figure of modernity. *History of the Human Sciences, 22*(1), 5–36. doi:10.1177/0952695108099133.

Walter, W. G. (1953). The electroencephalographic development of children. In J. M. Tanner & B. Inhelder (Eds.), *Discussions on child development* (Vol. 1, pp. 132–140). London: Tavistock Publications.

Summary

Taking care of oneself is increasingly interpreted as taking care of one's brain. Apart from drugs like antidepressants or ADHD medicines, there are many more options to stimulate the brain. Brain products vary from books, food, soft drinks, puzzles, toys, and games to—the topic of this book—brain devices. Without undergoing any surgery, and without seeing a doctor, people can, for example, try to change their brain frequencies with light and sound machines. They can also use devices that work with electric or magnetic stimulation, like cranial electrotherapy stimulation, transcranial direct current stimulation, or transcranial magnetic stimulation. Or they can try to change their brainwaves with a neurofeedback device that provides positive feedback whenever their brain produces the intended brainwave activity, for example, in the form of the movement of a racing car or a Pac-Man on a computer screen. These techniques are promoted for various psychotherapeutic uses as well as for self-enhancement, and sometimes also for spiritual purposes and mind-altering effects. They can be bought on the Internet, used in brain clinics, or people can try to build their own brain machines.

Using a brain device to cure or improve oneself can be described as a contemporary 'technology of the self', an expression Foucault used to

© The Editor(s) (if applicable) and The Author(s) 2016   **157**
J. Brenninkmeijer, *Neurotechnologies of the Self*,
DOI 10.1057/978-1-137-53386-9

refer to those techniques that 'permit individuals to effect by their own means or with the help of others a certain number of operations on their own bodies and souls, thoughts, conduct and way of being, so as to transform themselves in order to attain a certain state of happiness, purity, wisdom, perfection, or immortality' (Foucault, 1988, p. 18). Different techniques, Foucault explained, are based on different forms of care and constitute different modes of selves (Foucault, 1984, 1988). People can work on themselves, for example, by taking antidepressants, seeing a psychoanalyst, or confessing one's sins—and hence they will perceive themselves as persons with chemical unbalances, repressed sexual desires, or struggles with the devil, which are basically three different ways of being oneself. Following this idea of Foucault, this book presents a historical, ethnographic, and theoretical exploration of the mode of subjectivity that is constituted when people use a brain device to change themselves.

Chap. 2 gives an overview of the historical and contemporary uses of several brain technologies of the self, and analyzes why therapeutic brain devices do not have much scientific credibility yet. Following Ashmore, Brown, and Macmillan (2005), I argue that the scientific reliability of (therapeutic) effects partly depend on how findings are demonstrated, and for what public. Light and sound machines, for example, were often demonstrated with impressive histories—starting in prehistoric times, referring to recognizable and reproducible experiments, and full of famous spokespersons—but they were especially promoted as technologies for self-experimentation, and hence not so well presented in the scientific (public) domain. Electric and magnetic brain devices (like transcranial magnetic stimulation or transcranial direct current stimulation) are well demonstrated—with impressive histories, theatrical performances, and a professional demarcation policy—but representatives still have problems defending the therapeutic efficacy of these devices against skepticism, and hence attaining scientific credibility. Neurofeedback was historically promoted as both a scientific and a spiritual practice and even today experts have problems translating their therapeutic findings into experimental settings, and hence to transfer their results from a personal (spiritual and self-help) into a scientific (public and polemic) domain.

Chap. 2 analyzes why therapeutic brain devices do not have much scientific credibility yet, but in the rest of the book it becomes clear that this lack of approval does not mean that these devices do not have effects. I use historical and ethnographic methods to find out how working on the self by working on the brain influences our way of being oneself. For this, I focus on neurofeedback, because this technique literally confronts people with their brainwave activity and directly asks them to intervene in this activity. To understand how brainwaves and selves got involved with each other, Chap. 3 explores the work and biographies of four central figures in the history of neurofeedback, and the explanations of contemporary neurofeedback practitioners. I show that the 'discoverer of the human EEG', the German psychiatrist and psychophysiologist Hans Berger (1873–1941), was driven by a personal and spiritual mission, but became increasingly frustrated about the complicated relationship between physical and psychical events. One of his followers, the British neurophysiologist and cyberneticist William Grey Walter (1910–1977), introduced the first brainwave stimulation technologies, and hence the first struggles between brains and subjects. Moreover, he connected brainwaves to personality types by writing about brain brothers and strangers, and he entangled humans and machines by building robots with self-recognition and by describing the brain as an adaptive system. Two so-called 'founding fathers of neurofeedback', the American psychologists Joe Kamiya (1925–) and Barry Sterman (1935–), both performed experiments in which their subjects learned to train their brain at will, in order to become more spiritual, or to improve one's personality.

That is, Chap. 3 demonstrates that entanglements between brains and selves, humans and machines, and material and spiritual beliefs emerged in the work of early scientists, and are also present in the explanations of contemporary neurofeedback experts. Neurofeedback practitioners on the one hand claim that neurofeedback gives control back to the client, but on the other hand use language that comes close to the terminology of Walter, for example, when talking about 'low levels [of dopamine] walking into your door'. Moreover, in line with the ideas of neurofeedback pioneers, some contemporary practitioners combine their materialistic brain therapies with spiritual practices, such as meditation or yoga.

Chap. 4 investigates how contemporary neurofeedback users work on themselves by training their brain. My ethnographic research among clients and practitioners shows that neurofeedback therapy can change people's notion of themselves. Trying to change your brain activity with the purpose of improving yourself suggests that this self is, or is in, the brain. When explaining their actions, however, users make a clear distinction between their selves and their brains, with statements such as 'It seems that at the moment you start focusing, your brain interrupts.' Apparently, people start doing neurofeedback because they want to improve themselves, but to react to the feedback of their brains they have to distance their selves from their brains. In other words, the self is *extended* with the brain instead of coinciding with it.

Besides a self and a brain, other entities can also be identified. Neurofeedback users sometimes designate their subconscious, their will power, the computer, or the practitioner as the actor that trains their brain. When explaining their problems, they refer to various brain-related entities like 'alpha and theta things' or 'yellow spot'. In addition to these brain-related entities, users involve their lives and psychology with statements like 'it is my life that made me quiet'. They describe themselves and the neurofeedback process in a rather computerized way ('my system resets itself'), and they often combine this materialistic view with spiritual practices like yoga or meditation. That is to say, users of neurofeedback constitute a new way of being themselves. This self can be described as an extended assembled or multiplied self made up of all kinds of entities that emerged in the neurofeedback process.

This change of the self should not be considered as only a matter of perception. According to neurofeedback users, the confrontation with the biological equivalent of their behavior gave them, for example, the experience of loss ('I had always thought I was controlled by myself'); of fear ('a bio-organic robot'); and of relief ('you lack a certain substance in your brain'). Some users encourage their friends or family members to do neurofeedback too. They often claim that they would prefer resuming sessions if their problems returned, and sometimes clients buy the equipment to train themselves (or their relatives), or to set up their own clinics. That is to say, doing neurofeedback does not only change the way people see themselves, their problems, and their responsibility, it also changes their behavior, their relationships, and the way they handle new problems. The new way of seeing themselves because of the neurofeedback has created a new way of being themselves. It is important to stress this ontological change since it demonstrates the effects of neurofeedback: irrespective of the clinical results, the effect of neurofeedback is very clear in the sense that it creates a new way of being oneself.

To understand how neurofeedback users constitute a new way of being, it is also necessary to analyze how other actors help them in creating this mode of self. As could already be observed in the interviews with users, 'the others' are not only humans (practitioners, scientists, or relatives), but also non-humans (computers, brain maps, games). Hence, Chap. 6 is based on the work of the sociologist of science Andrew Pickering who describes scientific practices as dances of agency between humans and non-humans. I show that not only the act of doing neurofeedback—trying to control your brainwaves, for example, by effecting a movement in a computer game—can be described as a dance of agency, but that the whole process involves various actors.

To become clients, people first have to be aware of their problems, their brains, and the 'solution'. Actors varying from relatives, psychiatrists, psychological tests, brain maps, neurofeedback practitioners, metaphors, and success stories are involved in this awareness process. During the therapy, practitioners use tools like footstools, neurofeedback teddy bears, or movies to keep their clients' attention. They do not only work with their clients' brains, but also have to calm down their minds and posture their bodies. Moreover, when questioning practitioners about the neurofeedback process many more actors become involved, while it is not always clear who or what is the leading agent. Practitioners are not always the experts because they sometimes simply lack the knowledge, protocols do not always work, quantitative electroencephalographs (qEEGs) can behave unpredictably, and computers sometimes appear to be the actors that define the training. As a result of these uncertainties, practitioners more or less experiment on their clients to find out which protocol, method, or frequency range works for which client. That is to say, neurofeedback can be described as a process of trial-and-error tinkering (Pickering, 1995), in which human and non-human actors perform a dance of agency, without following a clear choreography.

The envisioned result of this all is a restored or improved client. To recognize this improvement, however, practitioners often have to help their clients by pointing out what is changed, not only in their clients' brains, but mostly in their feelings, performances, or lives. For this, they appeal to many actors. Actors varying from parents, brain maps, to specific results can make the client conscious of his or her changed state, which can be found 'inside' of the brain (normalized brainwaves), mind (a reduction of stress), body (physical experiences), or somewhere outside in the world (tests, performances). Another kind of result, however, is again a dance of agency, but now concerning the self of the client. Some of the actors that emerged during the neurofeedback process—brainwaves, computers, colored spots—keep on working on the self of the client.

That is to say, this book demonstrates that using a device to understand, control, or cure the self by the brain can change people's notion of their selves. Contemporary neurofeedback users—clients and practitioners—make a blend of technical, physiological, spiritual, and personal statements to express themselves and how they deal with their problems. This 'new' way of self is personal and unique, but it is also a result of a historical quest to grasp the (spiritual) self with a brain device. I argue that trying to explain or improve the self by working on the brain constitutes an extended way of being oneself. Since this argument might disagree with many individual and scientific experiences, I finish with a reflection on my explanation of the extended self. I conclude that this self is complex, not because it is an activity of multiple entities, but because it contradicts common understandings of being a self. If we want to slow down this process of extending, for example, because it makes the self increasingly complicated to understand, feel free about, or responsible for, it is worthwhile to start making it visible.

References

Ashmore, M., Brown, S. D., & Macmillan, K. (2005). Lost in the mall with Mesmer and Wundt: Demarcations and demonstrations in the psychologies. *Science, Technology and Human Values, 30*(1), 76–110. doi:10.1177/0162243904270716.

Electronic patterns of the brain. (1956). *Life, 40*(15), 89–90.

Foucault, M. (1984). On the genealogy of ethics: An overview of work in progress. In P. Rabinow (Ed.), *The Foucault reader* (pp. 340–372). New York: Pantheon Books.

Foucault, M. (1988). Technologies of the self. In L. M. Martin, H. Gutman, & P. H. Hutton (Eds.), *Technologies of the self. A seminar with Michel Foucault* (pp. 16–49). Amherst: The University of Massachusetts Press.

Pickering, A. (1995). *The mangle of practice: Time, agency, and science.* Chicago/London: University of Chicago Press.

Appendix: Users

Practitioners

1. Neurofeedback researcher (Male), Professor, Europe
2. Practitioner and experimenter of neurofeedback experiments (Female), MSc Psychology, UK
3. Neuropsychologist (F), neurofeedback practitioner, course supervisor, PhD Psychology, UK
4. Computer expert (M), neurofeedback practitioner, course supervisor, UK
5. Neurofeedback practitioner (M), researcher, MSc Psychology (doing his PhD), the Netherlands
6. Neurofeedback practitioner (M), MSc Psychology, NL
7. Neurofeedback practitioner (M), PhD Psychology, NL
8. Neurofeedback practitioner (F), uses a different method of neurofeedback called Zengar, NL
9. Neurofeedback practitioner (M), researcher, PhD psychology, NL
10. Neurofeedback practitioner (F), course taker, UK
11. Neuropsychologist (M), course taker, PhD, UK

© The Editor(s) (if applicable) and The Author(s) 2016   **163**
J. Brenninkmeijer, *Neurotechnologies of the Self*,
DOI 10.1057/978-1-137-53386-9

29. Participant neurofeedback course (F), wants to train her child who is diagnosed with ADHD, UK

All practitioners use or used neurofeedback on themselves. Practitioners 2, 4, 5, 6, and 9 do or did this as a form of self-treatment, by themselves or with the help of another practitioner. Practitioners 2, 4, and 6 had no clear success so far.

Clients (All Dutch)

12. Client (M), multiple problems, went to several neurofeedback clinics, bought his own equipment, and regularly uses neurofeedback at home. At the moment he uses nutrition instead of neurofeedback
13. Client (M), was (successfully) treated for a depression, returned after some time to revitalize his training
14. Client (M), diagnosed with ADHD, went to several practitioners, switched to Zen meditation
15. Client (F), diagnosed with AD(H)D, went to two practitioners, quit because of money
16. Respondent to open questionnaire (M), burnout
17. Respondent (F), emotional and sleeping problems, stress
18. Respondent (F), multiple problems
19. Respondent (F), ADHD/depression
20. Respondent (M), stress and concentration problems
21. Respondent (M), motoric problems (spasm)
22. Respondent (F), panic, anxiety, and dissociative problems
23. Respondent (F), anxiety, stress
24. Respondent (M), burnout, fatigue
25. Respondent (M), ADHD
26. Respondent (F), anxiety, anorexia nervosa
27. Respondent (M), tinnitus
28. Respondent (M), hyperactivity

Most clients claimed they noticed positive effects from the neurofeedback. Three persons (19, 24, 26) did not notice any effects, or not yet, and for one person (27) the problem became worse.

Index

Note: Page number followed by 'n' refers to footnotes.

© The Editor(s) (if applicable) and The Author(s) 2016
J. Brenninkmeijer, *Neurotechnologies of the Self,*
DOI 10.1057/978-1-137-53386-9

The manufacturer's authorised representative in the EU is Springer
Nature Customer Service Centre GmbH, Europaplatz 3, 69115 Heidelberg,
Germany. If you have any concerns regarding our products, please
contact ProductSafety@springernature.com

Printed and bound by CPI Group (UK) Ltd, Croydon, CR0 4YY
23/04/2026
02095587-0013